Computational Solutions to

Practical Probability Problems

Computational Solutions to
Practical Probability Problems

Paul J. Nahin

PRINCETON UNIVERISTY PRESS

PRINCETON AND OXFORD

Copyright © 2008 by Princeton University Press
Published by Princeton University Press,
41 William Street,
Princeton, New Jersey 08540
In the United Kingdom:
Princeton University Press,
6 Oxford Street,
Woodstock OX20 1TW

Libarary of Congress Control Number 2007060123

ISBN-13: 978-0-691-12698-2

British Library Cataloging-in-Publication Data is available

This book has been composed in ITC New Baskerville

Printed on acid-free paper. ∞

pup.princeton.edu

Printed in the United States of America

10 9 8 7 6 5 4 3 2 1

To the memory of

Victor Benedict Hassing (1916–1980)

who taught me mathematics at the "old" Brea-Olinda Union High School, Brea, California (1954–58) and who, when he occasionally walked to school with me in the morning, would enthusiastically tell me of his latest mathematical reading

Comments on Probability and Monte Carlo

An asteroid or comet impact is the only natural disaster that can wipe out human society.... A large impact is an improbable event that is absolutely guaranteed to occur. Over the span of geological time, very large impacts have happened countless times, and will occur countless more times in ages to come. Yet in any given year, or in one person's lifetime, the chance of a large impact is vanishingly small. The same, it should be noted, was also true when the dinosaurs ruled Earth. Then, on one ordinary day, probability arrived in the form of a comet, and their world ended.
—Curt Pebbles, *Asteroids: A History* (Smithsonian Institution Press 2000), illustrating how even extremely-low-probability events become virtually certain events if one just waits long enough

Monte Carlo is the unsophisticated mathematician's friend. It doesn't take any mathematical training to understand and use it.
—MIT Professor Billy E. Goetz, writing with perhaps just a bit too much enthusiasm in *Management Technology* (January 1960)

The truth is that random events can make or break us. It is more comforting to believe in the power of hard work and merit than to think probability reigns not only in the casino but in daily life.
—Richard Friedman, M.D., writing in the *New York Times* (April 26, 2005) on mathematics in medicine

Analytical results may be hard to come by for these cases; however, they can all be handled easily by simulation.
—Alan Levine, writing in *Mathematics Magazine* in 1986 about a class of probability problems (see Problem 13 in this book)

The only relevant difference between the elementary arithmetic on which the Court relies and the elementary probability theory [of the case in hand] is that calculation in the latter can't be done on one's fingers.
—Supreme Court Justice John Harlan, in a 1971 opinion *Whitcomb v. Chevis*

Contents

MATLAB Solutions To The Problems   101

Computational Solutions to

Practical Probability Problems

Introduction

Three times he dropped a shot so close to the boat that the men at the oars must have been wet by the splashes—each shot deserved to be a hit, he knew, but the incalculable residuum of variables in powder and ball and gun made it a matter of chance just where the ball fell in a circle of fifty yards radius, however well aimed.
—from C. S. Forester's 1939 novel *Flying Colours* (p. 227),
   Part III of *Captain Horatio Hornblower*, the tale of a military man
   who understands probability

This book is directed to three distinct audiences that may also enjoy some overlap: teachers of either probability or computer science looking for supplementary material for use in their classes, students in those classes looking for additional study examples, and aficionados of recreational mathematics looking for what I hope are entertaining and educational discussions of intriguing probability problems from "real life." In my first book of probability problems, *Duelling Idiots and Other Probability Puzzlers* (Princeton University Press, 2000), the problems were mostly of a whimsical nature. Not always, but nearly so. In this book, the first test a problem had to pass to be included was to be practical, i.e., to be from some aspect of "everyday real life."

This is a subjective determination, of course, and I can only hope I have been reasonably successful on that score.

From a historical viewpoint, the nature of this book follows in a long line of precedents. The very first applications of probability were mid-seventeenth-century analyses of games of chance using cards and dice, and what could be more everyday life than that? Then came applications in actuarial problems (e.g., calculation of the value of an annuity), and additional serious "practical" applications of probabilistic reasoning, of a judicial nature, can be found in the three-century-old doctoral dissertation ("The Use of the Art of Conjecturing in Law") that Nikolaus Bernoulli (1687–1759) submitted to the law faculty of the University of Basel, Switzerland, in 1709.

This book emphasizes, more than does *Duelling Idiots*, an extremely important issue that arises in most of the problems here, that of *algorithm development*—that is, the task of determining, from a possibly vague word statement of a problem, just what it is that we are going to calculate. This is nontrivial! But not all is changed in this book, as the philosophical theme remains that of *Duelling Idiots*:

1. No matter how smart you are, there will always be probabilistic problems that are too hard for you to solve analytically.
2. Despite (1), if you know a good scientific programming language that incorporates a random number generator (and if it is good it will), you may still be able to get numerical answers to those "too hard" problems.

The problems in this book, and my discussions of them, elaborate on this two-step theme, in that most of them are "solved" with a so-called *Monte Carlo simulation*.[1] (To maximize the challenge of the book, I've placed all of the solutions in the second half—look there only if you're stumped or to check your answers!) If a theoretical solution does happen to be available, I've then either shown it as well—if it is short and easy—or provided citations to the literature so that you can find a derivation yourself. In either case, the theoretical solution can then be used to validate the simulation. And, of course, that approach can be turned on its head, with the simulation results being used to numerically check a theoretical expression for special cases.

In this introductory section I'll give you examples of both uses of a Monte Carlo simulation.

But first, a few words about probability theory and computer programming. How much of each do you need to know? Well, more than you knew when you were born—there is no royal road to the material in this book! This is not a book on probability *theory*, and so I use the common language of probability without hesitation, expecting you to be either already familiar with it or willing to educate yourself if you're not. That is, you'll find that words like *expectation, random walk, binomial coefficient, variance, distribution function*, and *stochastic processes* are used with, at most, little explanation (but see the glossary at the end of the book) beyond that inherent in a particular problem statement. Here's an amusing little story that should provide you with a simple illustration of the level of sophistication I am assuming on your part. It appeared a few years ago in *The College Mathematics Journal* as an anecdote from a former math instructor at the U.S. Naval Academy in Annapolis:

> It seems that the Navy had a new surface-to-air missile that could shoot down an attacking aircraft with probability 1/3. Some top Navy officer then claimed that shooting off three such missiles at an attacking aircraft [presumably with the usual assumptions of independence] would surely destroy the attacker. [The instructor] asked his mathematics students to critique this officer's reasoning. One midshipman whipped out his calculator and declared "Let's see. The probability that the first missile does the job is 0.3333, same for the second and same again for the third. Adding these together, we get 0.9999, so the officer is wrong; there is still a small chance that the attacking aircraft survives unscathed." Just think [noted the instructor], that student might himself be a top U.S. navy officer [today], defending North America from attack.

If you find this tale funny because the student's analysis is so wrong as to be laughable—even though his conclusion was actually correct, in that the top Navy officer was, indeed, wrong—and if you know how to do the correct analysis,[2] then you are good to go for reading this book. (If you think about the math abilities of the politicians who make decisions about the viability of so-called anti-ballistic missile shields and the equally absent lack of analytic talent in some of the people who advise them, perhaps you will find this tale not at all funny but rather positively scary.[3])

The same level of expectation goes for the Monte Carlo codes presented in this book. I used MATLAB 7.3 when creating my programs, but I limited myself to using only simple variable assignment statements, the wonderful rand (which produces numbers uniformly distributed from 0 to 1), and the common if/else, for, and while control statements found in just about all popular scientific programming languages. My codes should therefore be easy to translate directly into your favorite language. For the most part I have avoided using MATLAB's powerful vector/matrix structure (even though that would greatly reduce simulation times) because such structure is not found in all other popular languages. MATLAB is an incredibly rich language, with a command for almost anything you might imagine. For example, in Problem 3 the technical problem of sorting a list of numbers comes up. MATLAB has a built-in sort command (called—is this a surprise?— sort), but I've elected to actually code a sorting algorithm for the Monte Carlo solution. I've done this because being able to write sort in a program is not equivalent to knowing how to code a sort. (However, when faced again in Problem 17 with doing a sort I *did* use sort—okay, I'm not always consistent!) In those rare cases where I do use some feature of MATLAB that I'm not sure will be clear by inspection, I have included some explanatory words.

Now, the most direct way to illustrate the philosophy of this book is to give you some examples. First, however, I should admit that I make no claim to having written the best, tightest, most incredibly elegant code that one could possibly imagine. In this book we are more interested in problem solving than we are in optimal MATLAB coding. I am about 99.99% sure that every code in this book works properly, but you may well be able to create even better, more efficient codes (one reviewer, a sophisticated programmer, called my codes "low level"—precisely my goal!). If so, well then, good for you! Okay, here we go.

For my first example, consider the following problem from Marilyn vos Savant's "Ask Marilyn" column in the Sunday newspaper supplement *Parade Magazine* (July 25, 2004):

> A clueless student faced a pop quiz: a list of the 24 Presidents of the 19th century and another list of their terms in office, but scrambled. The object was to match the President with the term. He had to guess every time. On average, how many did he guess correctly?

To this vos Savant added the words,

> Imagine that this scenario occurs 1000 times, readers. On average, how many matches (of the 24 possible) would a student guess correctly? Be sure to guess before looking at the answer below!

The "answer below" was simply "Only one!" Now that is indeed surprising—it's correct, too—but to just say that and nothing else certainly leaves the impression that it is all black magic rather than the result of logical mathematics.

This problem is actually an ancient one that can be traced back to the 1708 book *Essay d'analyse sur les jeux de hazard* (*Analyses of Games of Chance*), by the French probabilist Pierre Rémond de Montmort (1678–1719). In his book Montmort imagined drawing, one at a time, well-shuffled cards numbered 1 through 13, counting aloud at each draw: "$1, 2, 3, \ldots$." He then asked for the probability that no card would be drawn with a coincidence of its number and the number being announced. He didn't provide the answer in his book, and it wasn't until two years later, in a letter, that Montmort first gave the solution. Montmort's problem had a profound influence on the development of probability theory, and it attracted the attention of such illuminances as Johann Bernoulli (1667–1748), who was Nikolaus's uncle, and Leonhard Euler (1707–1783), who was Johann's student at Basel.

Vos Savant's test-guessing version of Montmort's problem is not well-defined. There are, in fact, at least three different methods the student could use to guess. What I suspect vos Savant was assuming is that the student would assign the terms in a one-to-one correspondence to the presidents. But that's not the only way to guess. For example, a student might reason as follows: If I follow vos Savant's approach, it is possible that I could get every single assignment wrong.[4] But if I select one of the terms (*any* one of the terms) at random, and assign that same term over and over to each of the twenty-four presidents, then I'm sure to get one right (and all the others wrong, of course). Or how about this method: for each guess the student just randomly assigns a term from all twenty-four possible terms to each of the twenty four presidents. That way, of course, some terms may never be assigned, and others may be assigned more than once. But guess what—the average number of correct matches with this method is still one. Now that's *really* surprising! And finally, there's yet one more astonishing feature to this problem, but I'll save it for later.

Suppose now that we have no idea how to attack this problem analytically. Well, not to worry, a Monte Carlo simulation will save the day for us. The idea behind such a simulation is simple enough. Instead of imagining a thousand students taking a test, let's imagine a million do, and for each student we simulate a random assignment of terms to the presidents by generating a random permutation of the integers 1 through 24. That is, the vector term will be such that term(j), $1 \leq j \leq 24$, will be an integer from 1 to 24, with each integer appearing exactly once as an element in term: term(j) will be the term assigned to president *j*. The correct term for president *j* is term *j*, and so if term(j) = j, then the student has guessed a correct pairing. With all that said, the code of guess.m should be clear (in MATLAB, codes are called m-files and a program name extension is always .m).

Lines 01 and 02 initialize the variables M (the length of the two lists being paired) and totalcorrect (the total number of correct pairings achieved after a million students have each taken the test). (The line numbers are included so that I can refer you to a specific line in the program; when actually typing a line of MATLAB code one does not include a line number as was done, for example, in BASIC. The semicolons at the end of lines 01 and 02 are included to suppress distracting screen printing of the variable values. When we do want to see the result of a MATLAB calculation, we simply omit the terminating semicolon.) Lines 03 and 12 define a for/end loop that cycles the code through a million tests, with lines 04 through 11 simulating an individual test. At the start of a test, line 04 initializes the variable correct—the number of correct pairings achieved on that test—to zero. Line 05 uses the built-in MATLAB command randperm(M) to generate a random permutation of the integers 1 through M (= 24), and lines 06 and 10 define a for/end loop that tests to see if the condition for one (or more) matches has been satisfied. At the completion of that loop, the value of correct is the number of correct pairings for that test. Line 11 updates the value of totalcorrect, and then another test is simulated. After the final, one-millionth test, line 13 computes the average number of correct pairings.

```
guess.m
01    M = 24;
02    totalcorrect = 0;
03    for k = 1:1000000
04        correct = 0;
05        term = randperm(M);
06        for j = 1:M
07            if term(j) == j
08                correct = correct + 1;
09            end
10        end
11        totalcorrect = totalcorrect + correct;
12    end
13    totalcorrect/1000000
```

When I ran guess.m, the code gave the result 0.999804, which is pretty close to the exact value of 1. But what is really surprising is that this result is independent of the particular value of M; that is, there is nothing special about the $M = 24$ case! For example, when I ran guess.m for $M = 5$, 10, and 43 (by simply changing line 01 in the obvious way), the average number of correct pairings was 0.999734, 1.002005, and 0.998922, respectively. It's too bad that vos Savant said nothing about this. Now, just for fun (and to check your understanding of the Monte Carlo idea), write a simulation that supports the claim I made earlier, that if the student simply selects at random, for each president, a term from the complete list of twenty four terms, then the average number of correct pairings is still one. You'll find a solution in Appendix 1.

Let me continue with a few more examples. Because I want to save the ones dealing with everyday concerns for the main body of this book, the ones I'll show you next are just slightly more abstract. And, I hope, these examples will illustrate how Monte Carlo simulations can contribute in doing "serious" work, too. Consider first, then, the following problem, originally posed as a challenge question in a 1955 issue of *The American Mathematical Monthly*. If a triangle is drawn "at random" inside an arbitrary rectangle, what is the probability that the triangle is obtuse? This is, admittedly, hardly a question from real life, but it will do here to illustrate my positions on probability theory, Monte Carlo simulation, and programming—and it is an interesting, if somewhat abstract, problem in what is called *geometric probability* (problems in which probabilities are associated with the lengths, areas, and volumes of various shapes; the classic example is the well-known Buffon needle problem, found in virtually all modern textbooks[5]). The 1955 problem is easy to understand but not so easy to analyze theoretically; it wasn't solved until 1970. To begin, we first need to elaborate just a bit on what the words "at random" and "arbitrary rectangle" mean.

Suppose we draw our rectangle such that one of the shorter sides lies on the positive x-axis, i.e., $0 \leq x \leq X$, while one of the longer sides lies on the positive y-axis, i.e., $0 \leq y \leq Y$. That is, we have a rectangle with dimensions X by Y. It should be intuitively clear that, whatever the answer to our problem is, it is what mathematicians call *scale invariant*, which means that if we scale both X and Y up (or down) by the same factor, the answer will not change. Thus, we lose no generality by simply taking the actual value of X and scaling it up (or down) to 1, and then scaling Y by the same factor. Let's say that when we scale Y this way we arrive at L; i.e., our new, scaled rectangle is 1 by L. Since we started by assuming $Y \geq X$, then $L \geq 1$. If $L = 1$, for example, our rectangle is actually a square. To draw a triangle "at random" in this scaled rectangle simply means to pick three *independent* points (x_1, y_1), (x_2, y_2), and (x_3, y_3) to be the vertices of the triangle such that the x_i are each selected from a uniform distribution over the interval $(0,1)$ and the y_i are each selected from a uniform distribution over the interval $(0,L)$.

For a triangle to be obtuse, you'll recall from high school geometry, means that it has an interior angle greater than 90°. There can, of course, be only one such angle in a triangle! So, to simulate this problem, what we need to do is generate a lot of random triangles inside our rectangle, check each triangle as we generate it for obtuseness, and keep track of how many are obtuse. That's the central question here—if we have the coordinates of the vertices of a triangle, how do we check for obtuseness? The law of cosines from trigonometry is the key. If we denote the three interior angles of a triangle by A, B, and C and the lengths of the sides opposite those angles by a, b, and c, respectively, then we have

$$a^2 = b^2 + c^2 - 2bc \cos(A),$$

$$b^2 = a^2 + c^2 - 2ac \cos(B),$$

$$c^2 = a^2 + b^2 - 2ab \cos(C).$$

Or,

$$\cos(A) = \frac{b^2 + c^2 - a^2}{2bc},$$

$$\cos(B) = \frac{a^2 + c^2 - b^2}{2ac},$$

$$\cos(C) = \frac{a^2 + b^2 - c^2}{2ab}.$$

Since the cosine of an acute angle, i.e., an angle in the interval $(0, 90°)$, is positive, while the cosine of an angle in the interval $(90°, 180°)$ is negative, we have the following test for an angle being obtuse: the sum of the squares of the lengths of the sides forming that angle in our triangle, minus the square of the length of the side opposite that angle, must be negative. That is, all we need to calculate are the numerators in the above cosine formulas. This immediately gives us an easy test for the obtuseness-or-not of a triangle: to be acute, i.e., to not be obtuse, all three interior angles must have positive cosines. The code obtuse.m uses this test on one million random triangles.

obtuse.m
```
01    S = 0;
02    L = 1;
03    for k = 1:1000000
04        for j = 1:3
05            r(j) = rand;
06        end
07        for j = 4:6
08            r(j) = L*rand;
09        end
10        d1 = (r(1) − r(2))^2 + (r(4) − r(5))^2;
11        d2 = (r(2) − r(3))^2 + (r(5) − r(6))^2;
12        d3 = (r(3) − r(1))^2 + (r(6) − r(4))^2;
13        if d1 < d2 + d3&d2 < d1 + d3&d3 < d1 + d2
14            obtusetriangle = 0;
15        else
16            obtusetriangle = 1;
17        end
18        S = S + obtusetriangle;
19    end
20    S/1000000
```

Lines 01 and 02 initialize the variables S, which is the running sum of the number of obtuse triangles generated at any given time, and L, the length of the longer side of the rectangle within which we will draw triangles. In line 02 we see L set to 1 (our rectangle is in fact a square), but we can set it to any value we wish, and later I'll show you the results for both L = 1 and L = 2. Lines 03 and 19 are the for/end loop that cycle the simulation through one million triangles. To understand lines 04 through 09, remember that the notation I'm using for the three points that are the vertices of each triangle is (x_1, y_1) (x_2, y_2), and (x_3, y_3), where the x_i are from a uniform distribution over 0 to 1 and the y_i are from a uniform distribution over 0 to L. So, in lines 04 to 06 we have a for/end loop that assigns random values to the x_i, i.e., $x_1 = r(1)$, $x_2 = r(2)$, and $x_3 = r(3)$, and in lines 07 to 09 we have a for/end loop that assigns random values to the y_i, i.e., $y_1 = r(4)$, $y_2 = r(5)$, and $y_3 = r(6)$. Lines 10, 11, and 12 use the r-vector to calculate the lengths (squared) of the three sides of the current triangle. (Think of a^2, b^2, and c^2 as represented by d_1, d_2, and d_3.) Lines 13 through 17 then apply, with an if/else/end loop, our test for obtuseness: when obtuse.m exits from this loop, the variable obtusetriangle will have been set to either 0 (the triangle is not obtuse) or 1 (the triangle is obtuse). Line 18 then updates S, and the next random triangle is then generated. After the one-millionth triangle has been simulated and evaluated for its obtusness, line 20 completes obtuse.m by calculating the probability of a random triangle being obtuse. When obtuse.m was run (for L = 1), it produced a value of 0.7247 for this probability, which I'll call $P(1)$, while for L = 2 the simulation gave a value of $0.7979 = P(2)$.

In 1970 this problem was solved analytically,[6] allowing us to see just how well obtuse.m has performed. That solution gives the theoretical values of

$$P(1) = \frac{97}{150} + \frac{\pi}{40} = 0.72520648\cdots$$

and

$$P(2) = \frac{1{,}199}{1{,}200} + \frac{13\pi}{128} - \frac{3}{4}\ln(2) = 0.79837429\cdots.$$

I think it fair to say that obtuse.m has done well! This does, however, lead (as does the simulation results from our first code guess.m) to an obvious question: What if we had used not a million but fewer simulations? Theoretical analyses of the underlying mathematics of the Monte Carlo technique show that the error of the method decreases as $N^{-1/2}$, where N is the number of simulations. So, going from $N = 100 = 10^2$ to $N = 10{,}000 = 10^4$ (an increase in N by a factor of 10^2) should reduce the error by a factor of about $\sqrt{10^2} = 10$.

To illustrate more exactly just what this means, take a look at Figure 1, where you'll see a unit square in the first quadrant, with an inscribed quarter-circle. The area of the square is 1, and the area of the circular section is $\frac{1}{4}\pi$. If we imagine randomly throwing darts at the square (none of which miss), i.e., if we pick points uniformly distributed

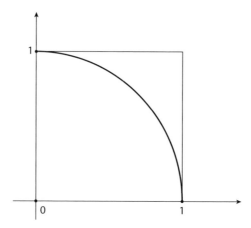

Figure 1. The geometry of a Monte Carlo estimation of π.

over the square, then we expect to see a fraction $\frac{(1/4)\pi}{1} = \frac{1}{4}\pi$ of them inside the circular section. So, if N denotes the total number of random points, and if P denotes the number of those points inside the circular section, then the fundamental idea behind the Monte Carlo method says we can write

$$\frac{P}{N} \approx \frac{1}{4}\pi,$$

or

$$\pi \approx \frac{4P}{N}.$$

We would expect this Monte Carlo estimate of π to get better and better as N increases. (This is an interesting use of a probabilistic technique to estimate a deterministic quantity; after all, what could be more deterministic than a constant, e.g., pi!)

The code pierror.m carries out this process for $N = 100$ points, over and over, for a total of 1,000 times, and each time stores the percentage error of the estimate arrived at for pi in the vector error. Then a histogram of the values in error is printed in the upper plot of Figure 2, thus giving us a visual indication of the distribution of the error we can expect in a simulation with 100 points. The width of the histogram

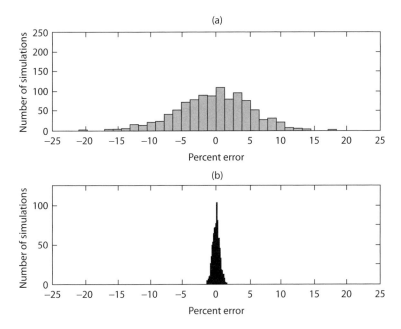

Figure 2. The error distribution in estimating π. a. Number of points per simulation = 100. b. Number of points per simulation = 10,000.

is a measure of what is called the variance of the error. The lower plot in Figure 2 shows what happens to the error variance when N is increased by a factor of 100, to 10,000 points per simulation. As mentioned earlier, theory says the error should decrease in this case by a factor of ten, and that is just what even a casual look at Figure 2 shows has happened. (I have not included in the code the MATLAB plotting and labeling commands that create Figure 2 from the vector error.) With high confidence we can say that, with $N = 100$ points, the error made by our Monte Carlo estimate of pi falls in the interval ±15%, and with high confidence we can say that, with $N = 10,000$ points, the error made in estimating pi falls in the interval ±1.5%. The theory of establishing what statisticians call a *confidence interval* can be made much more precise, but that would lead us away from the more pragmatic concerns of this book.

The lesson we'll take away from this is *the more simulations the better*, as long as our computational resources and available time aren't overwhelmed. A consequence of this is that Monte Carlo simulations

pierror.m

```
01      numberofpoints(1) = 100;
02      numberofpoints(2) = 10000;
03      for j = 1:2
04          N = numberofpoints(j);
05          for loop = 1:1000
06              P = 0;
07              for k = 1:N
08                  x = rand;
09                  y = rand;
10                  if x^2 + y^2 < 1
11                      P = P + 1;
12                  end
13              end
14              error(loop) = (4*P − pi*N)*100/(N*pi);
15          end
16      end
```

of rare events will require large values for N, which in turn requires a random number generator that is able to produce long sequences of "random" numbers before repeating.[7] This isn't to say, however, that running an ever longer simulation is the only way to reduce the statistical error in a Monte Carlo simulation. There are a number of other possibilities as well, falling into a category of techniques going under the general name of *variance reduction*; in Appendix 2 there is a discussion of one such technique. My general approach to convincing you that we have arrived at reasonably good results will be not nearly so sophisticated, however; I'll be happy enough if we can show that 10,000 simulations and 1,000,000 simulations give pretty nearly the same results. Mathematical purists may disagree with this philosophy, but this is not meant to be either a rigorous textbook or a theoretical dissertation. The variability of the estimates is due, of course, to different random numbers being used in each simulation. We could eliminate the variability (but not the error!) by starting each simulation with the random number generator seeded at the same point, but since your generator is almost surely different from mine, there seems no point in that. When you run one of my codes (in your favorite language) on your computer, expect to get pretty nearly the results reported here, but certainly don't expect to get the *same* results.

Let's try another geometric probability problem, this one of historical importance. In the April 1864 issue of *Educational Times*, the English mathematician J. J. Sylvester (1814–1897) submitted a question in geometric probability. As the English mathematician M. W. Crofton (1826–1915), Sylvester's one-time colleague and protégé at the Royal Military Academy in Woolwich, wrote in his 1885 *Encyclopaedia Britannica* article on probability,[8] "Historically, it would seem that the first question on [geometric] probability, since Buffon, was the remarkable four-point problem of Prof. Sylvester." Like most geometric probability questions, it is far easier to state than it is to solve: If we pick four points at random inside some given convex region **K**, what is the probability that the four points are the vertices of a concave quadrilateral? All sorts of different answers were received by the *Educational Times*: $\frac{1}{2}$, $\frac{1}{3}$, $\frac{1}{4}$, $\frac{3}{8}$, $\frac{35}{12\pi^2}$, and more. Of this, Sylvester wrote, "This problem does not admit of a deterministic solution." That isn't strictly so, as the variation in answers is due both to a dependency on **K** (which the different solvers had taken differently) and to the vagueness of what it means to say only that the four points are selected "at random." All of this variation was neatly wrapped up in a 1917 result derived by the Austrian-German mathematician Wilhelm Blaschke (1885–1962): If $P(\mathbf{K})$ is Sylvester's probability, then $\frac{35}{12\pi^2} \leq P(\mathbf{K}) \leq \frac{1}{3}$. That is, $0.29552 \leq P(\mathbf{K}) \leq 0.33333$, as **K** varies over all possible finite convex regions. For a given *shape* of **K**, however, it should be clear that the value of $P(\mathbf{K})$ is independent of the size of **K**; i.e., as with our first example, this is a scale-invariant problem.

It had far earlier (1865) been shown, by the British actuary Wesley Stoker Barker Woolhouse (1809–1893), that

$$P(\mathbf{K}) = \frac{4M(\mathbf{K})}{A(\mathbf{K})},$$

where $A(\mathbf{K})$ is the area of **K** and $M(\mathbf{K})$ is a constant unique to each **K**. Woolhouse later showed that $M(\mathbf{K}) = \frac{11A(\mathbf{K})}{144}$ and $M(\mathbf{K}) = \frac{289A(\mathbf{K})}{3,888}$ if **K** is a square or a regular hexagon, respectively. Thus, we have the results

$$\text{if } \mathbf{K} \text{ is a square, then } P(\mathbf{K}) = \frac{4 \times 11}{144} = \frac{11}{36} = 0.3055,$$

and

$$\text{if } \mathbf{K} \text{ is a regular hexagon, then } P(\mathbf{K}) = \frac{4 \times 289}{3,888} = \frac{289}{972} = 0.2973.$$

Both of these results were worked out by Woolhouse in 1867.

$M(\mathbf{K})$ was also known[9] for the case of $\mathbf{K}$ being a circle (a computation first done by, again, Woolhouse), but let's suppose that we don't know what it is and that we'll estimate $P(\mathbf{K})$ in this case with a computer simulation. To write a Monte Carlo simulation of Sylvester's four-point problem for a circle, we have two separate tasks to perform. First, we have to decide what it means to select each of the four points "at random" in the circle. Second, we have to figure out a way to determine if the quadrilateral formed by the four "random" points is concave or convex.

To show that there is indeed a decision to be made on how to select the four points, let me first demonstrate that there is indeed more than one way a reasonable person might attempt to define the process of random point selection. Since we know the problem is scale invariant, we lose no generality by assuming that $\mathbf{K}$ is the particular circle with unit radius centered on the origin. Then,

Method 1:   Let a "randomly selected" point have polar coordinates (r, θ), where r and θ are independent, uniformly distributed random variables over the intervals $(0,1)$ and $(0,2\pi)$, respectively.

Method 2:   Let a "randomly selected" point have rectangular coordinates (x, y), where x and y are independent, uniformly distributed random variables over the same interval $(0, 1)$ and such that $x^2 + y^2 \leq 1$.

The final condition in Method 2 is to ensure that no point is outside $\mathbf{K}$; any point that is will be rejected. Figure 3 shows 600 points selected "at random" by each of these two methods. By "at random" I think most people would demand a uniform distribution of the points over the

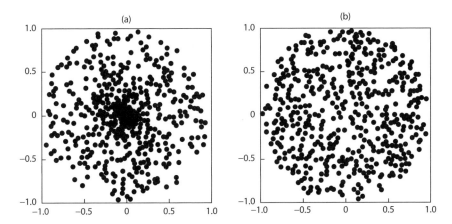

Figure 3. Two ways to generate points "at random" over a circle.
a. Method 1. b. Method 2.

area of the circle, and Figure 3 shows by inspection that this feature is present in Method 2 but is absent in Method 1 (notice the clumping of points near the center of the circle). What "at random" means was still a bit of a puzzle to many in the nineteenth century; Crofton wrote of it, in an 1868 paper in the *Philosophical Transactions of the Royal Society of London*, as follows:

> This [variation] arises, not from any inherent ambiguity in the subject matter, but from the weakness of the instrument employed; our undisciplined conceptions [that is, our intuitions] of a novel subject requiring to be repeatedly and patiently reviewed, tested, and corrected by the light of experience and comparison, before they [our intuitions, again] are purged from all latent error.

What Crofton and his fellow Victorian mathematicians would have given for a modern home computer that can create Figure 3 in a flash!

Method 2 is the way to properly generate points "at random," but it has the flaw of wasting computational effort generating many points that are then rejected for use (the ones that fail the $x^2 + y^2 \leq 1$ condition). Method 1 would be so much nicer to use, if we could eliminate the nonuniform clumping effect near the center of the circle.

This is, in fact, not hard to do once the reason for the clumping is identified. Since the points are uniformly distributed in the radial (r) direction, we see that a fraction r of the points fall inside a circle of radius r, i.e., inside a circle with area πr^2. That is, a fraction r of the points fall inside a smaller circle concentric with **K**, with an area r^2 as large as the area of **K**. For example, if we look at the smaller circle with radius one-half, then one-half of the points fall inside an area that is one-fourth the area of **K** and the other half of the points fall inside the annular region outside the smaller circle—a region that has an area three times that of the smaller circle! Hence the clumping effect near the center of **K**.

But now suppose that we make the radial distribution of the points vary not as directly with r, but rather as $\sqrt{r}$. Then a fraction r of the points fall inside a circle with area πr (remember, r itself is still uniform from 0 to 1), which is also a fraction r of the area of **K**. *Now* there is no clumping effect! So, our method for generating points "at random" is what I'll call Method 3:

Method 3:   Let r and θ be independent, uniformly distributed random variables over the intervals $(0,1)$ and $(0,2\pi)$, respectively. Then the rectangular coordinates of a point are $(\sqrt{r}\cos(\theta), \sqrt{r}\sin(\theta))$.

Figure 4 shows 600 points generated by Method 3, and we see that we have indeed succeeded in eliminating the clumping, as well as the wasteful computation of random points that we then would reject.

We are now ready to tackle our second task. Once we have four random points in **K**, how do we determine if they form a concave quadrilateral? To see how to do this, consider the so-called *convex hull* of a set of n points in a plane, which is defined to be the smallest convex polygon that encloses all of the points. A picturesque way to visualize the convex hull of a set of points is to imagine that, at each of the points, a slender rigid stick is erected. Then, a huge rubber band is stretched wide open, so wide that all the sticks are inside the rubber band. Finally, we let the rubber band snap tautly closed around the sticks. Those sticks (points) that the rubber band catches are the vertices of the convex hull (the boundary of the hull is the rubber band itself). Clearly, if our four points form a convex quadrilateral, then all

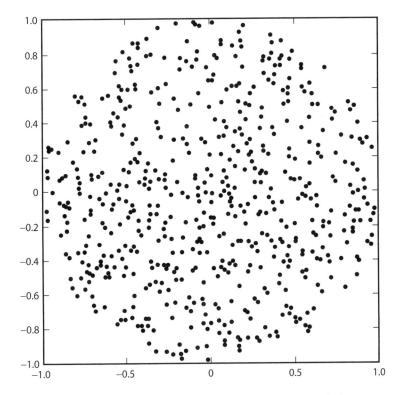

Figure 4. A third way to generate points "at random" over a circle.

four points catch the rubber band, but if the quadrilateral is concave, then one of the points will be inside the triangular convex hull defined by the other three points.

There are a number of general algorithms that computer scientists have developed to find the convex hull of n points in a plane, and MATLAB actually has a built-in function that implements one such algorithm.[10] So, this is one of those occasions where I'm going to tell you a little about MATLAB. Let $\mathbf{X}$ and $\mathbf{Y}$ each be vectors of length 4. Then we'll write the coordinates of our $n = 4$ points as $(\mathbf{X}(1), \mathbf{Y}(1))$, $(\mathbf{X}(2), \mathbf{Y}(2))$, $(\mathbf{X}(3), \mathbf{Y}(3))$, and $(\mathbf{X}(4), \mathbf{Y}(4))$. That is, the point with the "name" #k, $1 \leq k \leq 4$, is $(\mathbf{X}(k), \mathbf{Y}(k))$. Now, $\mathbf{C}$ is another vector, created from $\mathbf{X}$ and $\mathbf{Y}$, by the MATLAB function convhull; if we write $\mathbf{C}$ = convhull($\mathbf{X}$, $\mathbf{Y}$), then the elements of $\mathbf{C}$ are the names of the points

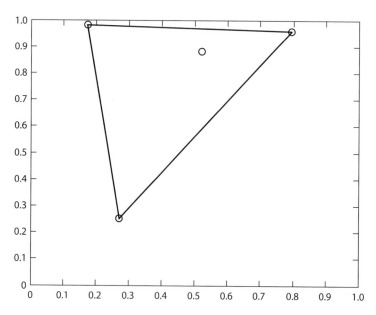

Figure 5. Convex hull of a concave quadrilateral.

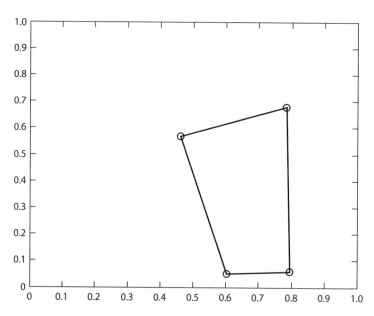

Figure 6. Convex hull of a convex quadrilateral.

on the convex hull. For example, if

$$\mathbf{X} = [0.7948 \quad 0.5226 \quad 0.1730 \quad 0.2714],$$

$$\mathbf{Y} = [0.9568 \quad 0.8801 \quad 0.9797 \quad 0.2523],$$

then

$$\mathbf{C} = [4 \quad 1 \quad 3 \quad 4],$$

where you'll notice that the first and last entry of $\mathbf{C}$ are the name of the same point. In this case, then, there are only three points on the hull (4, 1, and 3), and so the quadrilateral formed by all four points must be concave (take a look at Figure 5). If, as another example,

$$\mathbf{X} = [0.7833 \quad 0.4611 \quad 0.7942 \quad 0.6029],$$

$$\mathbf{Y} = [0.6808 \quad 0.5678 \quad 0.0592 \quad 0.0503],$$

then

$$\mathbf{C} = [4 \quad 3 \quad 1 \quad 2 \quad 4],$$

and the quadrilateral formed by all four points must be convex because all four points are on the hull (take a look at Figure 6).

We thus have an easy test to determine concavity (or not) of a quadrilateral: If $\mathbf{C}$ has four elements, then the quadrilateral is concave, but if $\mathbf{C}$ has five elements, then the quadrilateral is convex. The MATLAB function length gives us this information (length(v) = number of elements in the vector v), and it is used in the code sylvester.m, which generates one million random quadrilaterals and keeps track of the number of them that are concave. It is such a simple code that I think it explains itself. When sylvester.m was run it produced an estimate of $P(\mathbf{K} = \text{circle}) = 0.295557$; the theoretical value, computed by Woolhouse, is $\frac{35}{12\pi^2} = 0.295520$. Our Monte Carlo simulation has done quite well, indeed! (When run for just 10,000 simulations, sylvester.m's estimate was 0.2988.)

For the final examples of the style of this book, let me show you two problems that I'll first analyze theoretically, making some interesting arguments along the way, which we can then check by writing Monte Carlo simulations. This, you'll notice, is the reverse of the process we

---

sylvester.m

```
01      concave = 0;
02      constant = 2*pi;
03      for k = 1:1000000
04          for j = 1:4
05              number1 = sqrt(rand);
06              number2 = constant*rand;
07              X(j) = number1*cos(number2);
08              Y(j) = number1*sin(number2);
09          end
10          C = convhull(X,Y);
11          if length(C) == 4
12              concave = concave + 1;
13          end
14      end
15      concave/1000000
```

---

followed in the initial examples. Suppose, for the first of our final two problems, that we generate a sequence of independent random numbers x_i from a uniform distribution over the interval 0 to 1, and define the length of the sequence as L, where L is the number of x_i in the sequence until the first time the sequence fails to increase (including the first x_i that is less than the preceding x_i). For example, the sequence 0.1, 0.2, 0.3, 0.4, 0.35 has length $L = 5$, and the sequence 0.2, 0.1 has length $L = 2$. L is clearly an integer-valued random variable, with $L \geq 2$, and we wish to find its average (or *expected*) value, which we'll write as $E(L)$. Here's an analytical approach[11] to calculating $E(L)$.

The probability that length L is greater than k is

$$P(L > k) = P(x_1 < x_2 < x_3 < \cdots < x_k) = \frac{1}{k!}$$

since there are $k!$ equally likely permutations of the k x_i, only one of which is monotonic increasing. If $x_{k+1} < x_k$, then $L = k + 1$, and if $x_{k+1} > x_k$, then $L > k + 1$. In both cases, of course, $L > k$, just as claimed. Now, writing $P(L = k) = p_k$, we have by definition the answer to our question as

$$E(L) = \sum_{k=2}^{\infty} k p_k = 2p_2 + 3p_3 + 4p_4 + 5p_5 + \cdots,$$

which we can write in the form

$$
\begin{aligned}
E(L) = \quad & p_2 + p_3 + p_4 + p_5 + \cdots \\
+ \; & p_2 + p_3 + p_4 + p_5 + \cdots \\
+ \quad & \quad\; p_3 + p_4 + p_5 + \cdots \\
+ \quad & \quad\quad\;\; p_4 + p_5 + \cdots \\
+ \cdots & \, .
\end{aligned}
$$

The top two rows obviously sum to 1 (since all sequences have *some* length!). Thus,

$$E(L) = 2 + P(L > 2) + P(L > 3) + (P > 4) + \cdots = 2 + \frac{1}{2!} + \frac{1}{3!} + \frac{1}{4!} + \cdots.$$

But, since

$$e = \frac{1}{0!} + \frac{1}{1!} + \frac{1}{2!} + \frac{1}{3!} + \frac{1}{4!} + \cdots = 2 + \frac{1}{2!} + \frac{1}{3!} + \frac{1}{4!} + \cdots,$$

we see that $E(L) = 2.718281\cdots$. Or is it? Let's do a Monte Carlo simulation of the sequence process and see what that says. Take a look at the code called mono.m—I think its operation is pretty easy to follow. When run, mono.m produced the estimate $E(L) = 2.717536$, which is pretty close to the theoretical answer of e (simulation of 10,000 sequences gave an estimate of 2.7246).

This last example is a good illustration of how a lot of mathematics is done; somebody gets an interesting idea and does some experimentation. Another such illustration is provided with a 1922 result proven by the German-born mathematician Hans Rademacher (1892–1969): suppose t_k is $+1$ or -1 with equal probability; if $\sum_{k=1}^{\infty} c_k^2 < \infty$, then $\sum_{k=1}^{\infty} t_k c_k$ exists with probability 1 (which means it is possible for the sum to diverge, but that happens with probability zero, i.e., "hardly ever"). In particular, the so-called random harmonic series (RHS), $\sum_{k=1}^{\infty} \frac{t_k}{k}$, almost surely exists because $\sum_{k=1}^{\infty} \frac{1}{k^2}$ is finite. The question

mono.m
```
01    sum = 0;
02    for k = 1:1000000
03        L = 0;
04        max = 0;
05        stop = 0;
06        while stop == 0
07            x = rand;
08            if x > max
09                max = x;
10            else
11                stop = 1;
12            end
13            L = L + 1;
14        end
15        sum = sum + L;
16    end
17    sum/1000000
```

of the distribution of the sums of the RHS is then a natural one to ask, and a theoretical study of the random harmonic series was done in 1995. The author of that paper[12] wanted just a bit more convincing about his analytical results, however, and he wrote, "For additional evidence we turn to simulations of the sums." He calculated a histogram—using MATLAB—of 5,000 values of the partial sums $\sum_{k=1}^{100} \frac{t_k}{k}$, a calculation (agreeing quite nicely with the theoretical result) that I've redone (using instead 50,000 partial sums) with the code rhs.m, which you can find in Appendix 3 at the end of this book. I've put it there to give you a chance to do this first for yourself, just for fun and as another check on your understanding of the fundamental idea of Monte Carlo simulation (you'll find the MATLAB command hist very helpful—I used it in creating Figure 2—in doing this; see the solution to Problem 12 for an illustration of hist).

Now, for the final example of this Introduction, let me show you a problem that is more like the practical ones that will follow than like the theoretical ones just discussed. Imagine that a chess player, whom I'll call A, has been challenged to a curious sort of match by two of his competitors, whom I'll call B and C. A is challenged to play three sequential games, alternating between B and C, with the first game being with the player of A's choice. That is, A could play either BCB (I'll call this sequence 1) or CBC (and this will be sequence 2). From experience with B and C, A knows that C is the stronger player (the tougher for A to defeat). Indeed, from experience, A attaches probabilities p and q to his likelihood of winning any particular game against B and C, respectively, where $q < p$. The rule of this peculiar match is that to win the challenge (i.e., to win the match) A must win two games in a row. (This means, in particular, that even if A wins the first and the third games—two out of three games—A still loses the match!) So, which sequence should A choose to give himself the best chance of winning the match?

What makes this a problem of interest (besides the odd point I just mentioned) is that there are seemingly two different, indeed contradictory, ways for A to reason. A could argue, for example, that sequence 1 is the better choice because he plays C, his stronger opponent, only once. On the other hand, A could argue that sequence 1 is *not* a good choice because then he *has* to beat C in that single meeting in order to win two games in a row. With sequence 2, by contrast, he has two shots at C. So, which is it—sequence 1 or sequence 2?

Here's how to answer this question analytically. For A to win the match, there are just two ways to do so. Either he wins the first two games (and the third game is then irrelevant), or he loses the first game and wins the final two games. Let P_1 and P_2 be the probabilities A wins the match playing sequence 1 and sequence 2, respectively. Then, making the usual assumption of independence from game to game, we have

for sequence 1 : $P_1 = pq + (1-p)qp$

and

for sequence 2 : $P_2 = qp + (1-q)pq$.

Therefore

$$P_2 - P_1 = [qp + (1-q)pq] - [pq + (1-p)qp] = (1-q)pq - (1-p)qp$$
$$= pq[(1-q) - (1-p)] = pq(p-q) > 0$$

because $q < p$. So, sequence 2 always, for any $p > q$, gives the greater probability for A winning the challenge match, even though sequence 2 requires A to play his stronger opponent twice. This strikes most, at least initially, as nonintuitive, almost paradoxical, but that's what the math says. What would a Monte Carlo simulation say?

The code chess.m plays a million simulated three-game matches; actually, each match is played twice, once for each of the two sequences, using the same random numbers in each sequence. The code keeps track of how many times A wins a match with each sequence, and

---

chess.m

```
01    p = input('What is p?');
02    q = input('What is q?');
03    prob(1,1) = p;prob(1,3) = p;prob(2,2) = p;
04    prob(1,2) = q;prob(2,1) = q;prob(2,3) = q;
05    wonmatches = zeros(1,2);
06    for loop = 1:1000000
07       wongame = zeros(2,3);
08       for k = 1:3
09          result(k) = rand;
10       end
11       for game = 1:3
12          for sequence = 1:2
13             if result(game) < prob(sequence,game)
14                wongame(sequence,game) = 1;
15             end
16          end
17       end
18       for sequence = 1:2
19          if wongame(sequence,1) + wongame(sequence,2) == 2|...
                wongame(sequence,2) + wongame(sequence,3) == 2
20             wonmatches(sequence) = wonmatches(sequence)+1;
21          end
22       end
23    end
24    wonmatches/1000000
```

---

so arrives at its estimates for P_1 and P_2. To understand how the code works (after lines 01 and 02 bring in the values of p and q), it is necessary to explain the two entities prob and wongame, which are both 2×3 arrays. The first array is defined as follows: prob(j,k) is the probability A wins the kth game in sequence j, and these values are set in lines 03 and 04. Line 05 sets the values of the two-element row vector wonmatches to zero, i.e., at the start of chess.m wonmatches(1) = wonmatches(2) = 0, which are the initial number of matches won by A when playing sequence 1 and sequence 2, respectively. Lines 06 and 23 define the main loop, which executes one million pairs of three-game matches. At the start of each such simulation, line 07 initializes all three games, for each of the two sequences, to zero in wongame, indicating that A hasn't (not yet, anyway) won any of them. Then, in lines 08, 09, and 10, three random numbers are generated that will be compared to the entries in prob to determine which games in each of the two sequences A wins. This comparison is carried out in the three nested loops in lines 11 through 17, which sets the appropriate entry in wongame to 1 if A wins that game. Then, in lines 18 through 23, the code checks each row in wongame (row 1 is for sequence 1, and row 2 is for sequence 2) to see if A satisfied at least one of the two match winning conditions: winning the first two or the last two games. (The three periods at the end of the first line of line 19 is MATLAB's way of continuing a line too long to fit the width of a page.) If so, then line 20 credits a match win to A for the appropriate sequence. Finally, line 24 give chess.m's estimates of P_1 and P_2 after one million match simulations.

The following table compares the estimates of P_1 and P_2 produced by chess.m, for some selected values of p and q, to the numbers

		Theoretical		*Simulated*	
p	q	P_1	P_2	P_1	P_2
0.9	0.8	0.7920	0.8640	0.7929	0.8643
0.9	0.4	0.3960	0.5760	0.3956	0.5761
0.4	0.3	0.1920	0.2040	0.1924	0.2042
0.4	0.1	0.0640	0.0760	0.0637	0.0755

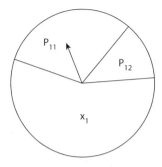
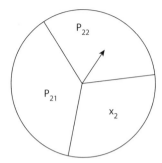

Figure 7. A spin game.

produced by the theoretical expressions we calculated earlier. You can
see that the simulation results are in excellent agreement with the
theoretical values.

Here's a little Monte Carlo challenge problem, of the same type as
the chess player's problem, for you to try your hand at (a solution is
given in Appendix 6). Consider the following game, which uses the
two spinner disks shown in Figure 7. Suppose a player spins one or
the other of the pointers on the disks according to the following rules:
(1) if the player spins pointer i and it stops in the region with area p_{ij},
he moves from disk i to disk j (i and j are either 1 or 2); (2) if a pointer
stops in the region with area x_i, the game ends; (3) if the game ends in
the region with area x_1, the player wins, but if the pointer stops in the
region with area x_2 the player loses. What is the probability the player,
starting with disk 1, wins? Assume the area of each disk is one, so that
$x_1 + p_{11} + p_{12} = 1$, as well as that $x_2 + p_{21} + p_{22} = 1$ (that is, all the x's
and p's are actually the probabilities that the pointer of either disk
stops in the respective region). This game can be analyzed theoretically
(which allows the code to be completely validated), and you'll find that
solution in Appendix 6, too. Run your code for the case of $p_{11} = 0.2$,
$p_{12} = 0.4$, $p_{21} = 0.3$, and $p_{22} = 0.35$.

I'll end this introduction—so we can get to the fun stuff of the
problems—with two quotations. The first is from from Professor Alan
Levine (see Problem 13), and it forms the central thesis of this book:

> From a pedagogical point of view, the problem and its solution
> as we have presented it here illustrate the fact that mathematics

is discovered in much the same way as any other science—by experimentation (here, simulation) followed by confirmation (proof). All too often, students think mathematics was created by divine inspiration since, by the time they see it in class, all the "dirty work" has been "cleaned up."

In this book, we'll be paying close attention to the "dirty work"!

The second quote is from the mathematical physicist W. Edwards Deming (1908–1993), and it makes a forceful statement on the value of being able to write computer codes like the ones in this book:

> If you can't describe what you are doing as a process [read *process* as *computer code*], you don't know what you are doing.

Right on!

References and Notes

1. Computer programs that use a random number generator to simulate a physical process are called *Monte Carlo codes*, in recognition of the famous gambling casino. This term was used by the pioneers of the method—Stanislaw Ulam (1909–1984), John von Neumann (1903–1957), and Nicholas Metropolis (1915–1999)—from the earliest days, and in his paper, "The Beginning of the Monte Carlo Method," published in *Los Alamos Science* (Special Issue, 1987, pp. 125–130), Metropolis reveals that *he* was the originator of the name. In a paper written some years later ("The Age of Computing: A Personal Memoir," *Daedalus*, Winter 1992, pp. 119–130), Metropolis wrote the astonishing words, "The Monte Carlo method ... was developed by Ulam and myself without any knowledge of statistics; to this day the theoretical statistician is unable to give a proper foundation to the method." I think the general view among mathematicians is that this last statement is not so. Computer scientists have introduced the general term of *randomized algorithm*, which includes Monte Carlo simulations of physical processes as a subset. A randomized algorithm is any algorithm (computer code) that uses a random number generator. In addition to Monte Carlo codes, there is now another category called *Las Vegas* codes (which are of no interest in this book). You can find more on all of this in a fascinating paper by Don Fallis, "The Reliability of Randomized Algorithms" (*British Journal for the Philosophy of Science*, June 2000, pp. 255–271).

2. I hope no reader thinks I am picking on Annapolis by repeating this tale (one that I suspect is quite popular at the United States Military Academy at West Point). I taught during the academic year 1981–1982 at the Naval Postgraduate School in Monterey, California, and many of my students were Annapolis graduates who were quite good at mathematics. The correct answer to the missile intercept problem is that if a missile successfully intercepts its target with probability 1/3, then of course it *misses* its target with probability 2/3. Three identical, independent missiles all fail to hit the same target, then, with probability $(2/3)^3 = 8/27$. So, at least one of the missiles does hit the target with probability $1 - (8/27) = 19/27 = 0.704$, considerably less than the certainty thought by the "top Navy officer." As the opening quotation makes clear, however, not all naval officers are ignorant of the inherent uncertainty of success in a three-missile attack on a target. An explanation of Hornblower's observation can be found in A. R. Hall, *Ballistics in the Seventeenth Century*, (Cambridge: Cambridge University Press, 1952, p. 55): "Nothing was uniform in spite of official efforts at standardisation; powder varied in strength from barrel to barrel by as much as twenty per cent; shot differed widely in weight, diameter, density and degree of roundness. The liberal allowances for windage, permitting the ball to take an ambiguous, bouncing path along the barrel of the gun, gave no security that the line-of-sight would be the line of flight, even if the cannon had been perfect. There was little chance of repeating a lucky shot since, as the gun recoiled over a bed of planks, it was impossible to return it to its previous position, while the platform upon which it was mounted subsided and disintegrated under the shock of each discharge."

3. Consider, for example, this quotation (*Washington Post*, July 22, 2005) from Air Force Lt. Gen. Henry Obering, director of the U.S. Missile Defense Agency: "We have a better than zero chance of successfully intercepting, I believe, an inbound warhead." This is no doubt true—but of course a high-flying eagle lost in a snowstorm with a mininuke clamped in its beak could make the same claim. After all, though "rather small," it is still true that $10^{-\text{googolplex}} > 0$.

4. The probability of that happening—of getting no pairings of presidents and terms correct—is 0.368, which is not insignificant. In general, the probability of getting m correct pairings when assigning M terms to M presidents, where the M terms are each uniquely assigned to a president, is given by $\frac{1}{m!}\sum_{k=0}^{M-m}\frac{(-1)^k}{k!}$. For $m = 0$ and $M = 24$ this formula gives the probability of zero correct pairings as very nearly $e^{-1} \approx 0.368$. See, for example, Emanuel Parzen, *Modern Probability Theory and Its Applications* (New York: John Wiley & Sons, 1960, pp. 77–79). You can find a scholarly, readable history of Montmort's problem, including detailed discussions of how the greats of yesteryear calculated their solutions to the problem, in L. Takács, "The Problem of Coincidences" (*Archive for History of Exact Sciences*,

21 [no. 3], 1980, pp. 229–244). The standard textbook derivation of the above formula uses combinatorial arguments and the inclusion-exclusion principle of probability. There is, however, a very clever way to numerically calculate the probability of no pairings, to any degree of accuracy desired, by using recursion. See Appendix 4 for how to do that, along with a MATLAB code that estimates the probability of no pairings with a Monte Carlo simulation. The inclusion-exclusion principle is discussed in Appendix 5, both theoretically and experimentally (i.e., it is illustrated with a simple MATLAB code).

5. In addition to one or more of those texts, you can find some interesting discussion on the Buffon needle problem in "Nineteenth-Century Developments in Geometric Probability: J. J. Sylvester, M. W. Crofton, J.-É. Barbier, and J. Bertrand" (*Archive for History of Exact Sciences*, 2001, pp. 501–524), by E. Senata, K. H. Parshall, and F. Jongmans.

6. Eric Langford, "A Problem in Geometric Probability" (*Mathematics Magazine*, November–December 1970, pp. 237–244). This paper gives the general solution for all $L \geq 1$, not just for the special cases of $L = 1$ and $L = 2$. The analysis in that paper inspired the following related question. If two points are independently and uniformly located in the unit interval, they divide that interval into three segments. What is the probability that those three segments form an obtuse triangle? You can find a theoretical analysis of this question in *Mathematics Magazine* (November–December 1973, pp. 294–295), where the answer is given as $\frac{9}{4} - 3\ln(2) = 0.170558\cdots$. The Monte Carlo code obtuse1.m produced the estimate 0.170567 using one million simulations of randomly dividing the unit interval into three parts. The variable S in line 01 is the same

---

obtuse1.m

```
01    S=0;
02    for k=1:1000000
03        point1=rand;
04        point2=rand;
05        if point1 > point2
06            temp=point1;
07            point1=point2;
08            point2=temp;
09        end
10        a=point1;
11        b=point2 - point1;
12        c=1 - point2;
13        if a+b > c&a+c > b&b+c > a
14            d1=a^2;
15            d2=b^2;
```

(continued)

(continued)

```
16          d3=c^2;
17          if d1 < d2+d3&d2 < d1+d3&d3 < d1+d2
18              obtusetriangle=0;
19          else
20              obtusetriangle=1;
21          end
22          S=S+obtusetriangle;
23      end
24  end
25  S/1000000
```

S as in obtuse.m. Lines 03 and 04 define the variables point1 and point2, and lines 05 through 09 ensure that their values are such that $0 < point1 < point2 < 1$. Thinking of point1 and point2 as the two points selected at random in the unit interval, then lines 10, 11, and 12 calculate the values of a, b, and c as the lengths of the three sides of a would-be triangle. Line 13 determines whether those sides do, in fact, satisfy the triangle inequalities that must be satisfied if and only if a triangle is possible (in a triangle, the sum of the lengths of any two sides is greater than the length of the remaining side); if they do, then the rest of the code is simply that of obtuse.m, which determines whether the triangle is an obtuse triangle.

7. For more on random number generators, see my book, *Duelling Idiots* (Princeton, N. J.: Princeton University Press, 2000, pp. 175–197). The generator in MATLAB 7.3, for example, will produce $2^{1492} > 10^{449}$ random numbers before repeating. If that generator had begun producing numbers at the prodigious rate of one trillion per second from the moment the universe was created (famously known as the Big Bang), about fifteen billion years ago, then it would have produced about 4.5×10^{29} numbers up to now. This is an infinitesimal fraction of the 7.3 generator's cycle length. To gain some historical appreciation of modern random number generators, consider the following complaint made by Lord Kelvin (Scottish engineer, physicist, and mathematician William Thomson [1824–1907]) in a lecture given in April 1900 at the Royal Institution of Great Britain (you can find his talk in the July 1901 issue of *The London, Edinburgh, and Dublin Philosophical Magazine and Journal of Science* under the title, "Nineteenth Century Clouds over the Dynamical Theory of Heat and Light"). When studying a problem in theoretical thermodynamics, Kelvin performed what may be the very first Monte Carlo simulation of a physical process. To select, with equal probability, from among 200 possibilities, he used 100 cards numbered from 0 to 99 (subjected before each drawing to a "very thorough shuffling") and coupled the result with the outcome of a coin toss. Alas, any real coin is almost certainly not

، fair, but Kelvin said nothing about how he accounted for that bias—this can be done for any coin; do you see how? The answer is in Appendix 7. In a footnote Kelvin says he also tried to replace the cards with small pieces of paper drawn from a bowl, but found "In using one's finger to mix dry [pieces] of paper, in a bowl, very considerable disturbance may be expected from electrification [i.e., from static electricity!]." Kelvin would have loved MATLAB's rand, but truly practical Monte Carlo had to wait until the invention of the high-speed electronic computer more than forty years later (as well as for advances in theoretical understanding on how to build random number generators in software). For a nice discussion on the modern need to achieve very high-speed generation of random numbers, see Aaldert Compagner, "Definitions of Randomness" (*American Journal of Physics*, August 1991, pp. 700–705).

8. For more on Crofton's contributions to geometric probability, see B. Eisenberg and R. Sullivan, "Crofton's Differential Equation" (*American Mathematical Monthly*, February 2000, pp. 129–139).

9. Richard E. Pfiefer, "The Historical Development of J. J. Sylvester's Four Point Problem" (*Mathematics Magazine*, December 1989, pp. 309–317).

10. So, unlike my argument about sort, I am not going to create the detailed code for implementing a convex hull algorithm. And why not, you ask? Well, I have to leave *something* for you to do! See, for example, R. L. Graham, "An Efficient Algorithm for Determining the Convex Hull of a Finite Planar Set" (*Information Processing Letters*, 1972, pp. 132–133).

11. Harris S. Shultz and Bill Leonard, "Unexpected Occurrences of the Number *e*" (*Mathematics Magazine*, October 1989, pp. 269–271). See also Frederick Solomon, "Monte Carlo Simulation of Infinite Series" (*Mathematics Magazine*, June 1991, pp. 188–196).

12. Kent E. Morrison, "Cosine Products, Fourier Transforms, and Random Sums" (*American Mathematical Monthly*, October 1995, pp. 716–724) and Byron Schmuland, "Random Harmonic Series" (*American Mathematical Monthly*, May 2003, pp. 407–416). The probabilistic harmonic series is interesting because the classic harmonic series, where $c_k = +1$, always, diverges. For more discussion on this and on the history of $\sum_{k=1}^{\infty} 1/k^2$, see my book, *An Imaginary Tale: The Story of $\sqrt{-1}$* (Princeton, N. J.: Princeton University Press, 1998, 2006 [corrected ed.], pp. 146–149).

The Problems

1. The Clumsy Dishwasher Problem

A broken dish is not something to take lightly. It was a broken, dirty banquet dish that killed the French mathematician Édouard Lucas in 1891; he died, at age fourty-nine, from an erysipelas infection resulting from a cut after a sharp fragment from a dish dropped by a waiter flew up and hit him in the face.

Suppose a restaurant employs five dishwashers. In a one-week interval they break five dishes, with four breakages due to the same individual. His colleagues thereafter call him "clumsy," but he claims it was just bad luck and could have happened to any one of them. The problem here is to see if he has some valid mathematical support for his position. First, see if you can calculate the probability that the same dishwasher breaks *at least* four of the five dishes that are broken (this includes, of course, the event of his breaking all five). It's an easy combinatorial calculation. Assume the dishwashers are equally skilled and have identical workloads, and that the breaking of a dish is a truly random event. If this probability is small, then the hypothesis that the given dishwasher actually *is* clumsy is more compelling than the hypothesis that a low-probability event has occurred. (What "low" means is, of course, subjective.) Second, after you have calculated this probability—and even if you can't—write a Monte Carlo simulation that estimates this probability. Are the two approaches in agreement?

2. Will Lil and Bill Meet at the Malt Shop?

Journeys end in lovers meeting...
—Shakespeare, *Twelfth Night*

The introduction used some examples of geometric probability from pure mathematics to open this book, and here's a problem from real life that can also be solved with a geometric probability approach. It can, however, also be easily attacked with a Monte Carlo simulation if the theoretical solution escapes you, and so I think it perfect for inclusion in this collection.

I used this problem in every undergraduate probability class I taught at the University of New Hampshire, and I originally thought the initial puzzlement I saw on my students' faces was because the math was strange to them. Then I learned it was mostly due to not knowing what a malt shop is—or I should say was. In New England, an ice cream milkshake is called a frappe, a term that, as a born-in-California boy, I had never heard before moving to New Hampshire in 1975. (Older readers can relive the nostalgia by watching Episode 189—"Tennis, Anyone?"—or Episode 195—"Untogetherness"—of *Leave It to Beaver* on the TV Land channel in which casual mentions of "the malt shop" are made. Younger readers, ask your grand parents!) So, if the malt shop is a problem for you, we could simply have Lil and Bill meeting instead at the theater, the high school gym, the local fast-food outlet, and so on, but as a fellow who went to high school in the 1950s, I still like the malt shop. Anyway, here's the problem.

Lil and Bill agree to meet at the malt shop sometime between 3:30 and 4 o'clock later that afternoon. They're pretty casual about details, however, because each knows that the other, while he or she will show up during that half-hour, is as likely to do so at any time during that half-hour as at any other time. If Lil arrives first, she'll wait five minutes for Bill, and then leave if he hasn't appeared by then. If Bill arrives first, however, he'll wait seven minutes for Lil before leaving if she hasn't appeared by then. Neither will wait past 4 o'clock. What's the probability that Lil and Bill meet? What's the probability of their meeting if Bill reduces his waiting time to match Lil's (i.e., if both waiting times are five minutes)? What's the probability of their meeting if Lil increases her waiting time to match Bill's (i.e., if both waiting times are seven minutes)?

3. A Parallel Parking Question

Starting with an n-dimensional random distribution of points, if each point is joined to its nearest neighbor, clusters are formed such that each point in a cluster is the nearest neighbor to another point in the cluster. I believe this may be a novel method of clustering with interesting applications in other fields—astrophysics and the traveling salesman problem, to name two.

—Daniel P. Shine, the creator of the original form of the following
    problem

Suppose $n \geq 2$ cars parallel park in a long, straight, narrow lot of length L. To make things easy, we'll imagine that the cars have no extensions but instead can be modeled as simply points distributed along a line segment of length L. We'll measure the parking location or *position* of a car by its distance from the lot entrance at the origin, and so each car is located in the interval $(0,L)$. Now, each car will obviously have a *nearest neighbor*. For example, if $n = 3$ and if the positions of car1, car2, and car3 are 0.05, 0.17, and 0.56, respectively, then the nearest neighbor of car2 is car1 (and, of course, the nearest neighbor—the only neighbor!—for car1 is car2; car1 and car2 are *mutual* nearest neighbors in this example). What is the probability that, if a car is picked at random, it is the nearest neighbor of the car that is *its* nearest neighbor? That is, if a car is picked at random, what is the probability it is one of a pair of *mutual* nearest neighbors?

For $n = 2$ cars, the answer is obviously one. For $n = 3$ cars, we see that car1 always has car2 as its nearest neighbor, and car3 always has car2 as its nearest neighbor. Car2 will have either car1 or car2 as its nearest neighbor. That is, either car1 and car2 are mutual nearest neighbors (and car3 is not part of a mutual nearest neighbor pair) or car2 and car3 are mutual nearest neighbors (and car1 is not part of a mutual nearest neighbor pair). In other words, for $n = 3$ there will always be precisely one mutual neighbor pair (involving two cars) and one lonely car that is not part of a pair. Thus, if we pick one of the three cars at random, it will be part of a mutual nearest neighbor pair with probability $\frac{2}{3}$. Now, what if $n = 4$ cars? If $n = 5$ cars? And so on.

Notice, whatever is the value of L, if after the n cars have parked we scale L by any positive factor, then the property of being a nearest neighbor is unaffected. That is, the property of neighborness is scale invariant. Thus, with no loss in generality, let's take $L = 1$. To simulate the parking of n cars at random in the lot, all we need do is to generate n numbers from a distribution uniform from 0 to 1, i.e., to make n calls to MATLAB's random number generator. Next, we sort these n numbers in increasing value, which is equivalent to determining the positions of the cars from left to right. We then simply proceed through these n numbers to determine how many are mutual nearest neighbors. Each increase in the mutual neighbor count means two more cars that are mutual nearest neighbors.

So, that's the problem: to write a Monte Carlo simulation to get estimates for the probabilities of a randomly selected car being part of a mutual nearest neighbor pair for the cases of $n = 4, 5, 6, \ldots, 12$, and for $n = 20$ and $n = 30$ cars. Do you notice anything interesting about the results? Are you, in fact, surprised by the simulation results? Can you guess what the probabilities are for any positive integer n? Can you derive those probabilities?

4. A Curious Coin-Flipping Game

*In 1876 Édouard Lucas (remember him from Problem 1?)
showed indirectly that $2^{67} - 1$ could not, as had been conjectured
since the early 1600s, be a prime number. Lucas could not,
however, actually factor this enormous number. Then, in 1903,
the American mathematician Frank Cole (1861–1926) gave a
remarkable paper at a meeting of the American Mathematical
Society. Without saying a word, he walked to the blackboard and
calculated $2^{67} - 1$. Then he multiplied out, longhand,*

$$193,707,721 \times 761,838,257,287.$$

*The two calculations agreed, and the audience erupted as one into
a thunderous, standing ovation. Cole later said it took him twenty
years of Sunday afternoons to factor*

$$2^{67} - 1 = 147,573,952,589,676,412,927.$$

Unsolved problems in mathematics are fun to play with because if
you manage to solve one, you'll become famous—or at least as famous
as mathematicians get to be. When Andrew Wiles (born 1953) finally
put Fermat's Last Theorem to rest in 1995, he became really famous,
but that sort of general fame is rare. Frank Cole's experience is more
typical; after his death, the American Mathematical Society named

a prize in his honor in 1928, but don't look for any breathless announcements about the yearly winner on your local newscast. (That important honor is reserved for the up-to-the-minute details on the latest Hollywood romance.) Almost as much fun as still unsolved problems are recently solved unsolved problems. They're fun because one can enjoy two simultaneous pleasures: (1) knowing the solution and (2) knowing that, until recently, nobody knew the solution. The best sort of this class of problem is one that is so easy to state that anybody, even a nonmathematician, can understand it, but until recently nobody could solve.

Here's an example of such a problem, first stated as a challenge question in the August–September 1941 issue of the *American Mathematical Monthly* (p. 483). It defied solution for a quarter-century, until 1966. As stated by its originator (G. W. Petrie, of the South Dakota State School of Mines):

> Three men have respectively l, m, and n coins which they match so that the odd man wins. In case all coins appear alike they repeat the throw. Find the average number of tosses required until one man is forced out of the game.

An unstated assumption is that all of the coins are *fair*; i.e., when flipped, all coins will show heads or tails with equal probability. The 1966 theoretical solution holds only for the case of fair coins. Now, just to be sure you have this game firmly in mind, here's a slight elaboration on Petrie's terse statement. When playing, each man selects one of his coins, and then all three simultaneously flip. If two coins show the same side and the third coin shows the opposite side, then the two men whose coins matched give those coins to the odd man out. If three heads or three tails is the outcome, however, nobody wins or loses on that toss. This process continues until one of the men loses his last coin. Notice that when you lose, you lose *one* coin, but when you win, you win *two* coins.

Write a Monte Carlo simulation that accepts values for l, m, and n, and then plays a large number of toss sequences, keeping track for each sequence of the number of tosses until one of the men goes broke (or, as mathematicians usually put it, is *ruined*). From that, the average number of tosses is easy to calculate. One advantage a simulation has over the theoretical analysis is that the *fair* assumption can be relaxed; more generally, we can write p for the probability that any coin will show heads, and the fair coin case of $p = 1/2$ becomes simply one special case. (One other significant plus for a simulation is that, even for just fair coins, there is no known theoretical answer for the case of more than three players; that, however, is a trivial extension for a simulation.) Write your code so that, along with the values of l, m, and n, the value of p is also an input parameter. To help you check your code, if $l = 1$, $m = 2$, and $n = 3$, then the theoretical answer (for $p = 1/2$, of course) is that the average length of a sequence until one man is ruined is two tosses. Make sure your code gives a result pretty close to that value. Now, what is the answer if $l = 2$, $m = 3$, and $n = 4$? If $l = m = n = 3$? If $l = 4$, $m = 7$, and $n = 9$? How do these answers change if $p = 0.4$?

5. The Gamow-Stern Elevator Puzzle

No opium-smoking in the elevators.
—sign in a New York City hotel (1907)

This is excellent advice, in general, and particularly so while studying the following problem, as you'll need all your wits unfogged if you are to have any chance of success.

Elevators are a wonderful source of fascinating probabilistic questions. They are easy to understand—elevators, after all, even though they are built to go straight up (with the exception of the fascinating elevators in the Luxor casino and hotel in Las Vegas)—aren't rocket science, and yet they present numerous situations in which the analyses are subtle and the results often nonintuitive. One such question that I knew would be in this book the moment I laid eyes on it is due to the mathematician Marvin Stern (1935–1974) and the physicist George Gamow (1904–1968). The problem originally appeared in a puzzle book by Gamow and Stern (*Puzzle-Math*, New York: Viking, 1958), but they did not have quite the correct theoretical analysis. So, let me quote from a paper[1] written by the computer scientist Donald Knuth that, more than a decade later, gave the correct solution to the problem:

> An amusing mathematical problem was devised by George Gamow and Marvin Stern, after they had been somewhat

frustrated by the elevator service in their office building. Gamow's office was on the second floor and Stern's on the sixth floor of a seven-story building. Gamow noted that, whenever he wished to visit Stern, the first elevator to arrive at the second floor was almost always "going down" not up. It seemed as though new elevators were being created at the top floor and destroyed at the ground floor, since no elevator ever would bypass the second floor intentionally on its way up. But when waiting for a descending elevator on the sixth floor, precisely the opposite effect was observed; the first elevator to pass was almost always "going up"!

To both Gamow and Stern it seemed almost as if there was a conspiracy to make them wait. In a world in which a conspiracy theory is put forth almost every day, in just about any imaginable setting, this is probably what many people would actually believe. There is, however, a perfectly logical mathematical explanation for what Gamow and Stern observed. The case of a building with just one elevator is easy to understand. We imagine that the elevator is continually running, going up and down all day long,[2] and so it seems reasonable to assume that, if Gamow requested its service at some arbitrary time, then with probability 1/6 it would be below his floor and with probability 5/6 it would be above his floor. Therefore, with probability 5/6 it would eventually arrive at his floor going down. For Stern it would be just the opposite, i.e., the elevator would, with probability 5/6, be going up when it arrived at his floor. This is what Gamow and Stern wrote and, so far so good. But then they blundered.

As Knuth wrote,

> When there is more than one elevator, Gamow and Stern say that "the situation will, of course, remain the same." But this is not true! Many a mathematician has fallen into a similar trap, being misled by something which seems self-evident, and nearly every example of faulty reasoning that has been published is accompanied by the phrase "of course" or its equivalent.

Knuth then quickly demonstrates that if there are two independent elevators, then the first elevator to arrive at Gamow's floor will be going down with probability 13/18, which is *not* equal to $5/6 = 15/18$. Check

Knuth's calculation with a Monte Carlo simulation. Also, what is the probability that the first-arriving elevator at Gamow's floor is going down in a three-elevator building?

References and Notes

1. Donald E. Knuth, "The Gamow-Stern Elevator Problem" (*Journal of Recreational Mathematics* 2, 1969, pp. 131–137).

2. As Knuth wrote, "Let us assume that we have an 'ideal' elevator system, which everyone knows does not exist, but which makes it possible to give a reasonable analysis. We will assume that each elevator goes continually up and down from the bottom floor to the top floor of the building, and back again in a cyclic fashion (independent of the other elevators). At the moment we begin to wait for an elevator on some given floor of the building [floor 2 for Gamow], we may assume that each elevator in the system is at a random point in its cycle, and that each will proceed at the same rate of speed until one [first] reaches our floor."

6. Steve's Elevator Problem

Much drinking, little thinking.
(So, no opium-smoking or boozing if you want to solve this problem!)

In March 2004 I received the following e-mail from a California reader of my first probability book, *Duelling Idiots*. That reader, Mr. Steve Saiz, wrote to ask for some help with an elevator problem different from the one in the last problem:

> Every day I ride up to the 15th floor in an elevator. This elevator only goes to floors G, 2, 8, 9, 10, 11, 12, 13, 14, 15, 16, and 17. On average, I noticed that I usually make 2 or 3 stops when going from the ground floor, G, to 15, but it really depends on the number of riders in the elevator car. Is it possible to find the expected value for the number of stops the elevator makes during my ride up to the 15th floor, given the number of riders?

Steve told me that he had written a Monte Carlo program in BASIC (using 10,000 simulations for each value of the number of riders), but he wasn't sure if it was correct. He assumed that the elevator stops are influenced only by the riders, and not by anyone on a floor waiting to get on. (This may, in fact, not be a bad assumption for Steve's situation—arriving at work in the early morning probably means there are few, if any, people on the upper floors waiting for elevator service.) For example, if Steve was alone on the elevator he would always experience just one stop on the way up to his floor. The first thing I did after reading Steve's e-mail was to write my own MATLAB simulation and check its results against the results produced by Steve's BASIC code. Here's that comparison, where my code used one million simulations for each value of the number of riders:

	Average number of stops	
Number of riders	*Steve's simulation*	*My simulation*
1	1	1
2	1.7295	1.727191
3	2.3927	2.388461
4	2.9930	2.989917
5	3.5457	3.535943

It seemed pretty likely from this comparison that Steve and I were getting the right numbers from our two independent simulations; but of course, the next obvious question is, is there an analytic expression for these numbers? If we let k denote the number of riders on the elevator in addition to Steve, then I was quickly able to work out such expressions for the $k = 1$ and $k = 2$ cases. Let me show you the $k = 1$ analysis, i.e., the case of two riders. To start, let's rephrase Steve's problem in a slightly more general way, as follows. There are n floors $(1, 2, 3, \ldots, n$, where $n = 11$ in Steve's problem) above floor G. One particular rider (Steve) always gets off on floor $n - 2$. Other riders, if present, can get off at any floor, each floor having probability $\frac{1}{n}$ as an exit floor. So, if there is one extra rider, i.e., if there are two riders on the elevator at floor G, then

there is *one* stop for Steve if the other rider gets off at Steve's floor *or above* (this happens with probability $\frac{3}{n}$)

or

there are *two* stops for Steve if the other rider gets off at any floor
below Steve's floor (this happens with probability $\frac{n-3}{n}$).

Thus, the average number of stops for Steve, when $k = 1$, is $1 \times \frac{3}{n} +$
$2 \times \frac{n-3}{n} = 2 - \frac{3}{n}$. For $n = 11$, this is $2 - \frac{3}{11} = \frac{19}{11} = 1.7273$, which agrees
nicely with the simulation results.

Here is the problem assignment for you. First, work out a theoretical
expression for the $k = 2$ case (there are three riders on the elevator at
floor G). Second, write a Monte Carlo simulation—you can check its
performance against the table of values I gave earlier—and use it to
extend the table to include the cases of $5 \leq k \leq 9$ (the cases of six to
ten riders, including Steve). As a special credit question, can you find
an exact theoretical expression for the average number of stops for
Steve, valid for any integer value of k? It can be done!

7. The Pipe Smoker's Discovery

It is better to light one candle than curse the darkness.
—Motto of the Christopher Society

(Yes, but first one needs to have a match.)

On one of the bookshelves next to my desk I have an old, well-thumbed 1970s textbook on computer programming which has, as one of its homework problems, the following interesting assignment:

> A pipe smoker has two booklets of matches in his pocket, each containing 40 matches initially. Whenever a match is required he picks one of the booklets at random, removing one match. Write a program using random numbers to simulate the situation 100 times and determine the average number of matches that can be removed until one booklet is completely empty.

I recall that the authors' answer at the back of the book (H. A. Maurer and M. R. Williams, *A Collection of Programming Problems and Techniques* [Prentice-Hall, 1972]) at first struck me as a bit strange—"between 61 and 79"—because the average value of a random quantity is a single number, not an interval. The answer as given is of course okay as the interval in which the average appears, and I soon realized that their answer was simply a helpful hint to the reader, giving some guidance as to the correct answer without giving the entire game away. So, what *is* the average number of matches removed until one booklet is empty? (Much progress in computer speed has been made since the 1970s, and so you can run a lot more than 100 simulations.)

This problem is not at all academic, and variations of it occur in unexpected places. See, for example, the next problem (A Toilet Paper Dilemma) for a generalization of enormous societal importance. Here's yet another one for you to think about right now, put forth some years ago by a physicist (I'll give you the citation in the solution): "To avoid running out of dental floss unexpectedly while traveling, I put two identical boxes, each containing 150 ft of floss, in my travel bag. I choose between the boxes randomly, and use 1 ft of floss each time. When one box becomes empty, how much floss can I expect to find in the other?" Matches, floss, toilet paper—, it's the same problem.

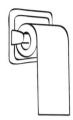

8. A Toilet Paper Dilemma

No job is over until all the paperwork is done.
—Graffiti written on at least half of all college dorm bathroom stalls
   in America

Up to now all the problems in this book have been studied with the aid of Monte Carlo simulations run on a computer. Not all probability problems that a computer is useful on use the Monte Carlo approach, however, and this problem is an example of that. A quarter-century ago Donald Knuth (look back at the Gamow-Stern elevator problem) published a paper in the *American Mathematical Monthly* that has achieved minor cult status among mathematicians and computer scientists. It analyzed a problem suggested to him by the architect of a new computer science building at Stanford University, and the paper[1] opens with a stunning flourish:

> The toilet paper dispensers in a certain building are designed to hold two roles of tissues, and a person can use either roll. There are two kinds of people who use the rest rooms in the building: *big-choosers* and *little-choosers*. A big-chooser always takes a piece of toilet paper from the roll that is currently larger; a little-chooser does the opposite. However, when the two rolls are the same size, or when only one roll is nonempty, everybody chooses the

nearest nonempty roll. When both rolls are empty, everybody has
a problem.

Well! Where in the world (I can imagine most subscribers to the
scholarly *American Mathematical Monthly* thinking when they read that)
is Knuth going with this? Pretty far, as it turns out.
    Knuth continues:

> Let us assume that people enter the toilet stalls independently at
> random, with probability p that they are big-choosers and with
> probability $q = 1 - p$ that they are little-choosers. If the janitor
> supplies a particular stall with two fresh rolls of toilet paper,
> both of length n, let $M_n(p)$ be the average number of portions
> left on one roll when the other roll empties. (We assume that
> everyone uses the same amount of paper, and that the lengths
> are expressed in terms of this unit.)

The similarity of this problem to the pipe smoker's problem is striking.
    Knuth's paper *analytically* studies the behavior of $M_n(p)$ for a fixed
p, as $n \to \infty$. Here, we'll be more interested in the behavior of $M_n(p)$,
$0 \le p \le 1$, for a fixed, finite n. (This problem can be attacked with
the Monte Carlo approach, but just for a change of pace we'll take a
different road here.) To this end, there are some things we can say
about $M_n(p)$ for any value of n. In his paper, for example, Knuth
writes, "It is easy to establish that $M_1(p) = 1$, $M_2(p) = 2 - p$, $M_3(p) =
3 - 2p - p^2 + p^3$, $M_n(0) = n$, $M_n(1) = 1$." The first and the last two, of
these statements are, in fact, actually obvious (with just a little thought).
For the first, we start with two rolls, each of length 1, and the very first
user picks one and so immediately empties that roll. The other roll
obviously still has length 1. QED.
    The penultimate statement is equally obvious as well, as $p = 0$
means *all* the users are little-choosers. The first user picks a roll and
uses it (thereby making it the little roll), and all subsequent users pick
it as well by definition (remember, $p = 0$) and run its length down to
zero. The other roll is never picked during that process and so is left
with all of its original n pieces as the other roll empties. QED.
    The last statement is only slightly more complicated to establish.
To start, we notice that $p = 1$ means *all* the users are big-choosers.

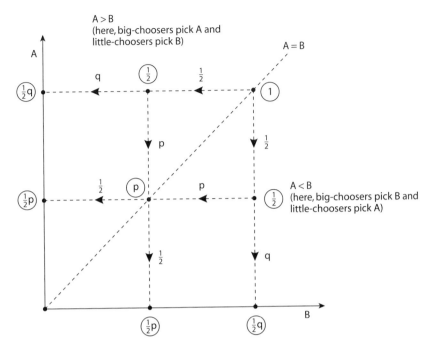

Figure P8.1. Calculating $M_2(p)$.

The first such user picks a roll and uses it (thereby making it the little roll). The next user therefore picks the *other* (big) roll, thereby making both rolls equal in length $(n - 1)$ once again. This back-and-forth process continues, which means that eventually we'll arrive at the state of two rolls with equal length 1. The next user then picks a roll and empties it, thereby leaving the other roll with length 1. QED.

What about the other expressions, for $M_2(p)$ and $M_3(p)$? Where do *they* come from? Knuth doesn't say explicitly, but it seems clear from his paper that it might have been as follows. Call the two rolls A and B and, as indicated by the first-quadrant lattice points (those points with integer-valued coordinates) in Figure P8.1, points on the diagonal line $A = B$ represent rolls of equal length. Points above the diagonal are all the different ways $A > B$ (that is, A is the big roll and B is the little roll), and points below the diagonal are all the different ways $A < B$. M_2 is for the case where we start at the upper right lattice point at $(2, 2)$, and users then drive the rolls from lattice point to lattice point until the state of the system reaches either the A-axis ($B = 0$) or the B-axis ($A = 0$), where one of the rolls is empty. Along each dashed line showing the possibility of a transition from one lattice point to the next I've written the probability of that transition, and next to each lattice point I've written (in a circle) the probability of being at that lattice point. If lattice point x can transition to lattice point x', then the probability of point x' is the probability of point x times the transition probability. From the figure we can now immediately write

$$M_2(p) = 2 \times \left(\tfrac{1}{2}q\right) + 1 \times \left(\tfrac{1}{2}p\right) + 1 \times \left(\tfrac{1}{2}p\right) + 2 \times \left(\tfrac{1}{2}q\right)$$

$$= q + \tfrac{1}{2}p + \tfrac{1}{2}p + q$$

$$= 2q + p = 2(1-p) + p = 2 - 2p + p = 2 - p,$$

as given by Knuth.

You'll find, if you repeat the above analysis for $M_3(p)$, that you'll arrive at Knuth's expression once again, but only after somewhat more algebra. Indeed, the algebraic demands with this approach go up pretty fast as the n in $M_n(p)$ increases, and this approach would be quite unattractive for computing, say, $M_{100}(p)$. For something like that, we need a different approach. The key idea is provided by Knuth. Following him, let's make a seemingly innocuous extension of our notation, writing $M_{m,n}(p)$ as the average number of portions remaining on one roll when the other roll empties, given that we start with m portions on one roll and n portions on the other. Then we immediately have, by definition, the statement

(a) $M_{n,n}(p) = M_n(p)$.

A second statement that is trivially obvious is

(b) $M_{n,0}(p) = n$.

Two additional statements are

(c) $M_{n,n}(p) = M_{n,n-1}(p), \quad n > 0$.

and

$$\text{(d) } M_{m,n}(p) = p M_{m-1,n}(p) + q M_{m,n-1}(p), \quad m > n > 0.$$

To establish (c), reason as follows: Given that we are at the diagonal lattice point (n, n), then with probability $1/2$ we move to lattice point $(n, n-1)$ and with probability $1/2$ we move to lattice point $(n-1, n)$. Thus,

$$M_{n,n}(p) = \frac{1}{2} M_{n,n-1}(p) + \frac{1}{2} M_{n-1,n}(p) = \frac{1}{2}[M_{n,n-1}(p) + M_{n-1,n}(p)].$$

But physically, $M_{n,n-1}(p) = M_{n-1,n}(p)$—in both expressions we are starting with two rolls, one of length n and the other of length $n-1$—and so we immediately have (c):

$$M_{n,n}(p) = M_{n,n-1}(p).$$

To understand (d), simply observe that, with p the probability a user is a big-chooser, we go from lattice point (m, n) to lattice point $(m-1, n)$ with probability p, and to lattice point $(m, n-1)$ with probability $q = 1 - p$. The constraints $m > n > 0$ mean that the starting lattice point (m, n) is *below* the diagonal. Now, why do we even care about these statements? Because, as Knuth writes, "The value of $M_n(p)$ can be computed *for all* n [my emphasis] from these recurrence relations, since no pairs (m', n') with $m' < n'$ will arise." Because of this we can study the behavior of $M_n(p)$ even if we can't find a direct formula for $M_n(p)$.

As an example of such a calculation, let's use the four recurrences to calculate $M_3(p)$. We have

$$M_3 = M_{3,3} = M_{3,2} = p M_{2,2} + q M_{3,1}$$

or, as $M_{2,2} = M_2 = 2 - p$, then

$$\begin{aligned} M_{3,3} &= p(2-p) + q[p M_{2,1} + q M_{3,0}] = p(2-p) + qp M_{2,1} + 3q^2 \\ &= p(2-p) + (1-p)p[p M_{1,1} + q M_{2,0}] + 3(1-p)^2 \\ &= p(2-p) + (1-p)p[p M_{1,0} + 2q] + 3(1-p)^2 \\ &= p(2-p) + (1-p)p[p + 2(1-p)] + 3(1-p)^2 \end{aligned}$$

$$= p(2-p) + (1-p)p[p+2-2p] + 3(1-p)^2$$
$$= 2p - p^2 + (1-p)p(2-p) + 3(1-p)^2$$
$$= 2p - p^2 + p(2-3p+p^2) + 3 - 6p + 3p^2$$
$$= 2p - p^2 + 2p - 3p^2 + p^3 + 3 - 6p + 3p^2$$
$$= 3 - 2p - p^2 + p^3,$$

which is Knuth's expression that I gave earlier.

Okay, at last, here's your problem. Use Knuth's recurrences to calculate the numerical values of $M_{200}(p)$ as p varies from 0 to 1, and plot $M_{200}(p)$. (In his paper, Knuth gives the plot for $M_{100}(p)$, but that seems like small rolls for the industrial-strength bathroom stalls you'd expect to find in the computer science building of a major[2] university!) Do you notice anything interesting happening around $p = 0.5$?

References and Notes

1. Donald E. Knuth, "The Toilet Paper Problem" *(American Mathematical Monthly*, October 1984, pp. 465–470.)

2. I hope readers understand I am not making fun of Stanford—I am, after all, a member of the Stanford class of 1962. It was at Stanford, in 1960, that I first learned computer programming—machine language programming—of the IBM 650, a terrifying experience that later allowed me to instantly appreciate modern computer languages.

9. The Forgetful Burglar Problem

Were it not better to forget
Than but remember and regret?
—from Letitia Elizabeth Landon's *Despondency*, with the following
   problem as a counter-example.

Imagine an old-time medicine-man show, wandering from town to town, selling narcotic-laced bottles of "Doctor Good" to the local hicks. Imagine further that the star of the show indulges in his own product just a bit too much for his own good, and so can't remember where he has already been in his travels. If he should put his tent up in a previously visited town, there may be a painful penalty extracted by enraged former customers who have since discovered the inability of narcotics to cure poor eyesight, arthritis, and thinning hair. A similar concern exists for a burglar who wanders up and down a street of homes plying his trade but who can't remember which houses he has already robbed. A return to a previously burgled home may result in a confrontation with an angry homeowner—who may be an armed and lethal member of the NRA!

For both our drunk medicine man and forgetful burglar, we are led to the following interesting mathematical question, first posed in 1958: Starting from an arbitrary town (home), if our medicine man (burglar) wanders up and down along an infinity of towns (homes) that are spaced uniformly along a line, taking with equal probability steps of either one or two towns (homes) in either direction (also with equal probability) to arrive at his next location, how long can our man get away with his crooked trade? That is, how long is it before a previously hoodwinked town or a previously burgled home is revisited? This is a very difficult problem to attack analytically, but it's all duck soup for a Monte Carlo simulation. In particular, find the probabilities that either of the men end up revisiting an old location on the kth step, where $1 \leq k \leq 7$. (The answer for the first case of $k = 1$ is, of course, trivial! Right?)

10. The Umbrella Quandary

Rain, rain, go away,
Come again another day.
—popular children's nursery rhyme

Imagine a man who walks every day between his home and his office. Because of the ever-present threat of rain, he likes to keep an umbrella at each location; that way, so goes his reasoning, if it is raining when he is about to leave one location to walk to the other, he won't get wet. The only flaw in his grand plan is that if it isn't raining he invariably neglects to take an umbrella with him. If you think about this for about five seconds, you should see that this can easily result in one location eventually having both umbrellas and the other location having none. If he then happens to be about to leave the no-umbrella location when it is raining, well, he's going to get wet!

We are thus led to the following pretty problem. If it is raining with probability p at the time the man is about to start each of his walks, then, on average, how many times will he remain dry before experiencing his first soaking? To make the problem just a bit more general, suppose we start the man off at home with $x > 0$ umbrellas, and with $y > 0$ umbrellas at the office, where x and y are input parameters to a Monte Carlo simulation. Use your simulation to answer the above question for the two cases of $x = y = 1$ and $x = y = 2$, as p varies from 0.01 to 0.99. That covers the entire spectrum from a very dry place (it rains, on average, once every 100 walks) to a very wet place (it doesn't rain, on average, once every 100 walks). Do you see why we don't need to run a simulation for either of the two extreme values of $p = 0$ (it never rains) and of $p = 1$ (it always rains)? That is, are the answers for these two special cases obvious to you by inspection?

11. The Case of the Missing Senators

Every vote counts.
—platitude heard before every election (occasionally it's even true)

Every vote counts, so vote early and often.
—cynical response to the first quote

Imagine that the U.S. Senate is about to vote on an important bill. It is known that there are more for votes than there are against votes, and so if all 100 senators show up, then the bill will pass. Let's suppose there are A senators who are against the bill, and thus there are $100 - A$ senators who are for it, where $A < 50$. If $A = 49$, for example, it will be a close 51 to 49 vote to pass the bill—if all the senators vote (and they don't always do!). Suppose, in fact, that M of the senators miss the vote. We are to imagine that these M senators are absent for purely random reasons (traffic delays, illness, forgetfulness, etc.) having nothing to do with whether a senator is either for or against the bill. It is then possible for the vote to go the wrong way, i.e., for the senators who do show up to collectively defeat the bill. What is the probability of that happening?

To answer this question, write a Monte Carlo simulation that accepts the values of A and M as input parameters. As a partial check on your code, the theoretical answer for the particular case of $A = 49$ and $M = 3$ is $\frac{51}{396} = 0.12878787\ldots$. (In the solution, I'll show you how to calculate this.) Make sure your code gives an estimate pretty close to this value for these values of A and M. What is the probability for $A = 49$ and $M = 4$? For $A = 49$ and $M = 5$?

12. How Many Runners in a Marathon?

In the Middle East a few years ago I was given permission by Israeli military authorities to go through the entire Merkava Tank production line. At one time I asked how many Merkavas had been produced, and I was told that this information was classified. I found it amusing, because there was a serial number on each tank chassis.
—a former military officer, thus proving that history still has lessons
   to teach, as the problem below explains

In any war, it is always of value to one side to have good intelligence on the weapons resources of the other side.[1] During the Second World War, for example, Allied military planners eagerly searched for ways to accurately estimate the Axis production of tanks, aircraft, and numerous other weapons platforms. In the specific case of German tanks, a very clever way to do that was based on using either the stamped serial numbers or the gearbox markings on captured Mark I or Mark V tanks, respectively.[2] As a modern writer (my source for the above quotation) explained, however, this type of problem has far wider applications than simply counting tanks:[3]

Suppose we have a population of objects labelled 1, 2, 3, ..., N with [the value of] N unknown. From a random sample X_1, X_2,

$X_3, \ldots, X_n$ of size n without replacement from this population we consider how to estimate N. One may estimate the number of runners in a race [hence the name of this problem], taxicabs in a city, or concession booths at a fair, for instance, based upon seeing just a sample of these labelled items.

It can be shown, in fact, that if one looks at a large number of such samples, each of size n, then the average value (the *expected* value) of all the *maximum* X_i's taken from each sample is given by

$$E(\max X_i) = \frac{n(N+1)}{n+1}.$$

This is not difficult to derive,[4] but we'll take it as a given here. Remember, what we are after is an estimate for N, the actual (unknown) maximum value in the population (equivalent, for this problem, to the size of the population). Solving for N gives

$$N = \frac{n+1}{n} E(\max X_i) - 1,$$

and if we assume that $E(\max X_i)$ is approximated by the maximum value observed in the sample we have, then we can form an estimate of N. For example, suppose we record the jersey numbers on fifteen marathon runners as they pass us at some point along the race—the assumption is that the numbers are from 1 to N, and were handed out before the race to the runners at random—and that the biggest number we see is 105. Then our estimate for N is

$$N \approx \frac{16}{15} \times 105 - 1 = 112 - 1 = 111.$$

There are at least a couple of interesting observations we can make about our estimation formula. First, since the observed maximum value in any sample of size n must be at least n, then the estimate for N must be at least

$$\frac{n+1}{n} \times n - 1 = n,$$

which says that the estimate for N can never be less than the largest value actually observed. This may seem absurdly obvious, but in fact there are other estimation formulas that people have considered that do not have this property! And second, if we have a sample of size N, i.e., if we have looked at the entire population, then our estimate for N will actually be exact. This is because in such a situation we are certain to have actually observed the maximum value N, and so our estimate is

$$\frac{N+1}{N} \times N - 1 = N,$$

which again makes absurdly obvious sense. It is, of course, possible for our estimation formula to make big errors, too. For example, suppose $N = 800$ and $n = 5$. A perfectly possible sample is [1, 2, 3, 4, 5], which gives the grossly incorrect estimate for N of

$$\frac{5+1}{5} \times 5 - 1 = 5.$$

Another possible sample is [1, 2, 3, 4, 800], in which we (unknowingly) have observed the actual value of N. Our formula, however, estimates N as

$$\frac{5+1}{5} \times 800 - 1 = \frac{6}{5} \times 800 - 1 = 960 - 1 = 959,$$

a value significantly greater than $N = 800$.

We can experimentally see how well our formula works in practice with a Monte Carlo simulation. That is, you are to write a program that, first, randomly picks the value of N (let's say N can be any integer from 100 to perhaps 1,000 or so), and then asks for the value of the sample size n as a *percentage* of N. That is, if you answer this question with 5, that means the sample size is to be 5% of the value the program picked for N; e.g., if $N = 260$, then the sample size will be $0.05 \times 260 = 13$. It is important to keep in mind that while you of course know the sample size percentage, you do not know the actual sample size because you do not know the value of N. The program then generates, randomly, n different integers[5] in the interval 1 to N (the program does know N, of course!), determines the maximum of those integers, and then uses our above estimation formula to arrive at an estimated value for N. This estimate can then be compared with the actual value of N to determine how well our estimation formula has performed. Specifically, have your code compute the percentage error in its estimate for each of 10,000 simulations using a fixed sample size percentage of 2% and plot a histogram of those errors. Repeat for sample size percentages of 5%, 10%, and 20%. Do your histograms have any interesting features?

References and Notes

1. The American invasion of Iraq in May 2003, justified in large part by erroneous information on weapons capabilities, is the most recent (as I write) illustration of how steep can be the price paid for faulty intelligence.

2. Richard Ruggles and Henry Brodie, "Empirical Approach to Economic Intelligence in World War II"(*Journal of the American Statistical Association*, March 1947, pp. 72–91). See also David C. Flaspohler and Ann L. Dinkheller, "German Tanks: A Problem in Estimation" (*Mathematics Teacher*, November 1999, pp. 724–728).

3. Roger W. Johnson, "Estimating the Size of a Population" (*Teaching Statistics*, Summer 1994, pp. 50–52).

4. Saeed Ghahramani, *Fundamentals of Probability* (Upper Saddle River, N. J.: Prentice-Hall, 1996, pp. 146–148). Johnson (previous note) also derives $E(\max X_i)$, but in a more heuristic way.

5. This should remind you of the previous problem (The Case of the Missing Senators), in which we had M randomly selected different senators missing a vote. In the language of a statistician, we had there a problem in *sampling a population* (the 100 senators who could potentially show up for the vote) *without replacement* to determine the M senators who miss the vote. You might want to review the code used in that problem—missing.m—for how that was accomplished, but for this problem you might also want to consider an elegant alternative (A. C. Bebbington, "A Simple Method of Drawing a Sample Without Replacement," *Applied Statistics*, 1975, no. 1, p. 136). As described by Bebbington, to select n items without replacement from N items, one simply looks one after the other at each of the N items. As you proceed, "after considering whether to select [initially, with probability $\frac{n}{N}$] each [item], N is reduced by 1, and if that [item] was selected, n is also reduced by 1. The next [item considered] is then selected with the new probability." Bebbington then writes the all-important "It is easily shown that this scheme must result in exactly n [items] being selected, and that every possible sample of the required size has an equal chance of occurring."

13. A Police Patrol Problem

When constabulary duty's to be done,
The policeman's lot is not a happy one.
—*Pirates of Penzance*, Gilbert and Sullivan

Imagine a long, straight stretch of high-speed, two-lane road that is routinely monitored by the state police for accidents.[1] There are many different ways that we can imagine how this monitoring could be implemented. For example:

(a) A police patrol car sits on a side of the road at the midpoint of the road, waiting for a radio call request for aid at the scene of an accident;

(b) A police patrol car continuously travels up and down from one end of the road to the other end and back again, listening for a radio request for aid;

(c) Two patrol cars independently implement (b), with the closer car responding to a request for aid.

To be even more general, we can imagine at least two additional ways to complicate methods (b) and (c):

(1) The stretch of road is a divided road, with the lanes in each direction separated by a grassy median strip whichallows a

patrol car to immediately drive to the accident, even if the accident is in the other lane and even if the patrol car has to reverse direction;

(2) The two lanes are separated by a concrete barrier, which means a patrol car can change lanes only at the two ends of the road.

With these possibilities, we therefore have six possible patrol scenarios: (a) and (1), (a) and (2), (b) and (1), (b) and (2), (c) and (1), and (c) and (2).

With these multiple scenarios available, an obvious question to ask is, which is best? This is an important issue that state officials might be faced with in deciding both how a highway should be constructed and how it should be policed. Not so obvious, however, is what we should use as a measure of "goodness,"; i.e., what are our criteria for "best"? There can easily be more than one answer to this question! We could, for example, argue that the best scenario is the one that, on average, gets a police car to the accident in the shortest time (which is equivalent to saying it is the scenario for which the police car has, on average, the shortest distance to travel). This is probably the first criterion that most people think of when presented with this question. But that isn't the only reasonable way to define what is meant by best. Here's another way. Let's suppose there is a critical, maximum response time T, beyond which even minor injuries may well become serious ones. This is equivalent to saying there is some travel distance D, beyond which even minor injuries may well become serious ones. So, another way to define the 'best' scenario is to say it is the one that minimizes the probability that the travel distance is greater than D. To keep the problem easy, however, let's agree to limit ourselves to the first criterion; i.e., for us, here, best will mean the fastest response time (the shortest travel distance) to the accident for the responding patrol car.

To study the six possible patrol scenarios, we can write Monte Carlo simulations of each and evaluate them using the above measure of best. In setting up these simulations, let's suppose accidents can happen, at random, at any location along the road in either lane, with x denoting the location of the accident. (Let's agree to measure all distances from the left end of the stretch of road.) We can, with no loss

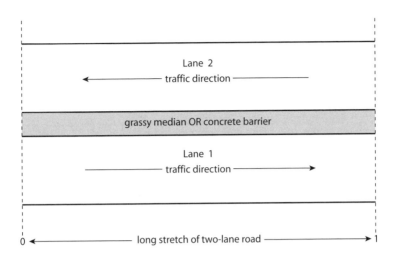

Figure P13.1. The geometry of the police patrol problem.

in generality, take the length of the road as unity, i.e., $0 \leq x \leq 1$. See Figure P13.1 for an illustration of all of this, where you'll notice I've labeled the lane with traffic flow to the right as Lane 1 and the lane with traffic flow to the left as Lane 2. Finally, we'll take the probability that an accident occurs in one lane or the other as equally likely.

Now, let's concentrate on a particular patrol car and write y as its distance from the left end of the road. Can you can see that there are eight distinct possibilities for the relative positions of the patrol car and the accident, taking into account the lanes that each is in? Specifically, if we define the direction of the patrol car to be either 0 or 1, depending on whether the patrol car is heading away from or toward the accident, respectively, at the instant the radio call for aid arrives, then the following table lists all eight possibilities. For each possibility the associated travel distance is given in the rightmost column for each of the two cases of interest, the grassy median between the two lanes and the concrete barrier between the two lanes. With this table in hand, we can now simulate each of the six patrol scenarios. Simply generate a random accident location (the value of x and the lane the accident is in) and a patrol car location (for (a) set $y = 1/2$—in either lane—it doesn't matter which, and for (b) use *rand* to set both the value of y and the lane the patrol car is in). Do this a million times, each

Patrol car lane	Accident lane	Direction	Distance to accident
1	1	0	$y - x$ for grass
			$2 + x - y$ for concrete
1	1	1	$x - y$
1	2	0	$y - x$ for grass
			$2 - x - y$ for concrete
1	2	1	$x - y$ for grass
			$2 - x - y$ for concrete
2	1	0	$x - y$ for grass
			$x + y$ for concrete
2	1	1	$y - x$ for grass
			$x + y$ for concrete
2	2	0	$x - y$ for grass
			$2 - x + y$ for concrete
2	2	1	$y - x$

time using the table to calculate the travel distance for the patrol car, both for a grassy median strip and a concrete barrier. To simulate (c), simply generate two (or more) randomly located patrol cars, calculate the travel distance to the accident for each car, and use the minimum of those values.

Your problem is to write the code that does all this for each of the six patrol scenarios, and then to determine which of the single car scenarios is, on average, best. Also, what is the impact of adding a second, random patrol car? Of adding a third or fourth car? As a final comment, I personally found this problem sufficiently busy with details that it was a great help, before starting to write code, to first construct a logic flow diagram. It is my understanding that the use of flow charts has fallen from grace with not just a few modern academic computer scientists, who consider their use to be akin to having to count on your toes to get beyond ten. If you feel that way, well, then, don't follow my suggestion. (But I really do think you'll find it helpful to draw a flow diagram.[2]) As a partial check on your code, it can be shown that for scenario (a) and (1) the average travel distance is 1/4, while for scenario (b) and (1) the average travel distance is 1/3. And for scenario (a) and (2), the average travel distance is 1. Use these special cases to partially validate the operation of your code.

References and Notes

1. This problem was inspired by the paper written by Alan Levine, "A Patrol Problem" (*Mathematics Magazine*, June 1986, pp. 159–166). I have, however, made some changes from Levine's original statement of the problem.

2. Two writers who agree with me are Andi Klein and Alexander Godunov, who write in their book *Introductory Computational Physics* (Cambridge: Cambridge University Press, 2006), in a section titled "Good Programming," that, before writing code, one should "design" a program and that "the best way to do that is with a flow chart." If you look at Appendix 6 (Solutions to the Spin Game) you'll see that I found it quite helpful to create a flow chart there, too, before beginning to write code. Many years ago, in the late 1960s, I worked for Hughes Aircraft Company in Southern California. During that employment I wrote the biggest program I've ever done, one that had over five thousand lines of assembly language code with real-time hardware/software interfaces between the computer and its magnetic tape drives and high-speed line printers. Believe me, I drew a *lot* of flow charts, for weeks, before I wrote even the first line of code!

14. Parrondo's Paradox

"Two wrongs don't make a right" is an old saying parents are fond of telling their children. And it's probably true, in a general sort of way, but there can be exceptions. Consider, for example, the following....

Here's another gambling problem based on coin flipping. While Problem 4 dated from 1941, this one is of much more recent vintage and is at least as challenging. It is, in fact, downright mystifying! Imagine that we have two games, called A and B, both based on flipping coins. Game A is quite easy to understand: you flip a so-called biased coin, i.e., a coin with unequal probabilities of showing heads and tails. Let's say that if the coin shows heads (with probability $1/2 - \epsilon$, $\epsilon > 0$), you win one dollar, otherwise you lose one dollar. It takes no great mathematical insight to appreciate that, on average, game A is a losing game. That is, if you play game A over and over, sometimes you'll win and the other times you'll lose, but you'll lose more than you win, and so, as you play more and more A games, your financial situation, as measured by your capital M, will worsen (M will decline).

Our other game is game B, with slightly more complicated rules: now you have two biased coins. If, at the time just before you select one of the coins to flip, your capital M is a multiple of three dollars, you will choose coin 1, which shows heads with probability $1/10 - \epsilon$. Otherwise you choose coin 2, which shows heads with probability $3/4 - \epsilon$. Again, as in game A, heads means you win one dollar and tails means you lose one dollar. It is not as obvious as it is with game A, but game B is a losing game, too. So, as with game A, if you play B over and over, your capital will tend to decline. This can be established analytically without too much trouble, but it is even easier to show by simply writing a Monte Carlo simulation, but don't do that—at least not yet!

Suppose now that you do the following: during a long sequence of games you randomly switch back and forth between playing loser game A and playing loser game B. It would seem obvious to most people, I think, that your capital M will definitely decline as you play. But is that really so? That question prompts your two assignments. First, write a Monte Carlo simulation of game B and thus demonstrate that B is, on average, a loser game as claimed. Second, write a Monte Carlo simulation of a gambler switching randomly between A and B during a long sequence of games, and plot the gambler's capital M as a function of time (imagine playing at the rate of one coin flip per second).

Since what we are studying here is a random process that evolves over time—what mathematicians call a *stochastic* process (the next problem, from queuing theory, is another example of a stochastic process)—we have to be careful about what we mean by talking of the average behavior of $M(k)$, where $k = 0, 1, 2, 3, \ldots$ is the number of coin flips the gambler has made. For example, if we imagine that "a long sequence of games" means 100 coin flips, then each time we play through such a sequence we'll almost certainly observe a different plot for $M(k)$ vs. k. Mathematicians call each such observation a *sample function* of the stochastic process. So, what your simulations of sequences of length 100 games should do is to play through many such sequences and then plot the average of the resulting sample functions, a calculation that produces what is called the *ensemble average* of the stochastic process (for the capital). Thus, if your codes simulate 10,000 sequences, each of length 100 games—a total of one million coin flips—and if the jth sequence produces the sample function $M_j(k)$, then the ensemble average that your codes should plot is $M(k)$ vs. k, where $M(k) = \frac{1}{10,000} \sum_{j=1}^{10,000} M_j(k)$, $1 \leq k \leq 100$. For both of your simulations use $\epsilon = 0.005$, assume the capital at $k = 0$ is always zero (negative capital means, of course, that you owe money), and in the second simulation, where you switch back and forth between the two games at random, play **A** and **B** with equal probability.

Are you surprised, perhaps even astonished, by the result from the second simulation? Nearly everyone is, and that's why this problem is called Parrondo's paradox, after the Spanish physicist Juan Parrondo, who discovered it in 1997.

15. How Long Is the Wait to Get the Potato Salad?

But the waiting time, my brothers,
Is the hardest time of all.
—*Psalms of Life* (1871) by Sarah Doudney

Probability theory has been a branch of mathematics for several hundred years, but new subspecialties are added on a fairly regular basis. About a hundred years ago the still active field of *queueing theory* was founded, which is the mathematical study of waiting. We all know what it is to have to wait for just about any sort of service, and examples of waiting are nearly endless. You have to wait, for example, or at least have to expect the possibility of waiting when you go into a haircutting salon, when you approach a checkout station at a grocery store, when you approach the toll booths on an interstate highway, when you wait for a washing machine in the hotel where you are staying on a trip, when you go to the DMV to apply for a driver's license, and on and on it goes.

The original motivation behind queueing theory was the development of the telephone exchange, with numerous subscribers asking for service from the exchange's finite processing resources (i.e., human operators). The mathematical questions raised by that technological development prompted the Danish mathematician Agner Erlang (1878–1929) to perform the first theoretical studies (published in 1909) of queues for the Copenhagen Telephone Company. Since Erlang's day, an enormous number of specific cases of queues have been analyzed, and the literature abounds with equations. For the problem here, however, we'll ignore all that and simply simulate the physics of a particularly common queue, the one that forms at most grocery store delicatessen counters.

As customers arrive at such counters, it is common practice to ask them to pull a numbered ticket out of a dispenser and then to wait as one or more clerks work their way through the numbers in sequential order.[1] Customers present themselves at the deli counter at an easily measured average rate (λ customers per hour, let's say), and the deli clerk takes various amounts of time to fill the various orders (of different sizes) of the customers. We'll take the service time for each customer to again be a random quantity, but as before, there will be an easily measured average rate of service (μ customers per hour, let's say). The store management will be interested in the answers to such mathematical questions as the following: (1) What is the average total time at the deli counter for the customers (total time is the sum of the waiting time and the service time)? (2) What is the *maximum* total time experienced by the unluckiest of the customers? (3) What is the average length of the customer waiting queue? and (4) What is the maximum length of the customer waiting queue? The answers to these questions are directly related to customer satisfaction and to the issue of how much physical space the store should expect to set aside for waiting customers to occupy. A fifth question is, What happens to the answers to the first four questions if a second, equally skilled deli clerk is hired? It should be obvious that the answers will all decrease, but more interesting is by how much? That answer will help store management decide if the second clerk is worth the additional expense in salary and benefits. An interesting sixth question might be, What is the fraction of the work day that the clerk(s) are idle (not serving a customer)?

The deli queue is a well-defined physical process, and so is an obvious candidate for a Monte Carlo simulation. Before we can do that, however, we need to define, carefully, what the phrases "random customer arrival" and "random customer service" mean. Suppose the deli opens for business at what we'll call time $t_0 = 0$. If customers thereafter arrive randomly at times t_1, t_2, t_3, ..., then the intervals between consecutive arrivals are $\Delta t_1 = t_1 - t_0$, $\Delta t_2 = t_2 - t_1$, $\Delta t_3 = t_3 - t_2$, and so on. Under very weak mathematical assumptions[2] (one of which is that more than one customer arriving at the same instant never happens), it turns out that these intervals are values of an exponentially distributed random variable (*not* a uniform one); i.e., if customers arrive at the average rate of λ per hour, then the Δt's are values of the random variable with the probability density function

$$\lambda e^{-\lambda t}, \quad t \geq 0$$

$$0, \quad t < 0.$$

In Appendix 8 you'll find a discussion on how to generate values for such a random variable from the values of a uniform random variable, and how to code the technique. In the same way, it is found that the values of the service times are the values of an exponential random variable with parameter μ; i.e., simply replace λ with μ in the above probability density function.

Write a Monte Carlo simulation of the deli queue over a single ten-hour business day (36,000 seconds) with the number of clerks (either one or two), λ, and μ as inputs. Notice that, in the case of a single deli clerk, there is the tacit assumption that $\mu > \lambda$; i.e., on average, the clerk can process customers faster than they arrive, else the queue length will tend to increase over time without limit, which would present an obvious problem for the store! This is more a mathematical than a practical concern, however, as we would see an unbounded queue only if the deli counter never closes. If the number of clerks is greater than one, however, it is mathematically possible to have $\mu < \lambda$ and still have a queue that would always remains finite. For each combination of values for λ, μ, and the number of clerks, run your code five times to observe the variation in the answers one might expect to see from day to day. In particular, answer the above six questions for the case of $\lambda = 30$ customers per hour and $\mu = 40$ customers per hour. Repeat for $\lambda = 30$ customers per hour and $\mu = 25$ customers per hour.

References and Notes

1. This is called a queue with a *first-come, first-served discipline*. For a deli, this seems like the obvious, fair way to do things (and it does indeed promote civility among even normally pushy customers), but it isn't the only possible discipline. For example, a hospital admitting desk might use a first-come, first-served discipline until a seriously ill or injured person arrived, and then that person would be given priority service even before all those who had arrived earlier were served. Yet another queue discipline was until recently routinely practiced by airlines when boarding passengers. Since passengers generally all board through the same cabin door at the front of the plane, it would seem logical to board passengers by their ticketed seating, with coach passengers at the rear of the plane boarding first, and then working forward to the front of the plane. And this was, in fact, what was done, with one exception. All the first-class passengers, who sit in the front of the plane, boarded first, and then the logical boarding order discipline was followed, generally causing much bumping, aisle congestion, and contorted maneuvering between the first-class passengers still in the aisle and coach passengers trying to get to the rear of the plane. First-class passengers should be boarded last, but apparently *last* is such a pejorative that it is thought, by the airline mental wizards who implement this goofy queue discipline, to be desirable. More recently, a number of airlines have adopted the discipline of boarding all window seat passengers "first" (but first-class passengers are still boarded *first*).

2. The technical name given by mathematicians to the random arrival of customers at our deli queue is that of *Poisson process*, after the French mathematician Simeon-Denis Poisson (1781–1840), who first (1837) described the probability density function of what we now call the Poisson random variable, which gives the probability of the total number of customers who have arrived at the deli by any time $t > 0$. The specific application of his mathematics to queues, however, came long after Poisson's death. Any good book on stochastic processes or operations research will develop the mathematics of Poisson queues.

16. The Appeals Court Paradox

How dreadful it is when the right judge judges wrong!
—Sophocles (495–405 B.C.)

Imagine a criminal appeals court consisting of five judges; let's call them A, B, C, D, and E. The judges meet regularly to vote (independently, of course) on the fate of prisoners who have petitioned for a review of their convictions. The result of each of the court's deliberations is determined by a simple majority; for a petitioner to be granted or denied a new trial requires three or more votes. Based on long-term record keeping, it is known that A votes correctly 95% of the time; i.e., when A votes to either uphold or to reverse the original conviction, he is wrong 5% of the time. Similarly, B, C, D, and E vote correctly 95%, 90%, 90%, and 80% of the time. (There are, of course, two different ways a judge can make a mistake. The judge may uphold a conviction, with new evidence later showing that the petitioner was in fact innocent. Or the judge may vote to reverse a conviction when in fact the petitioner is actually guilty, as determined by the result of a second conviction at the new trial.)

Write a Monte Carlo simulation of the court's deliberations, and use it to estimate the probability that the court, as an entity, makes an incorrect decision. (As a partial check on your code, make sure it gives the obvious answers for the cases of all five judges always being correct or always being wrong.) Then, change the code slightly to represent the fact that E no longer votes independently but rather now always votes as does A. Since A has a better (by far) voting record than does E, it would seem logical to conclude that the probability the court is in error would decrease. Is that what your simulation actually predicts?

17. Waiting for Buses

All things come to him who will but wait.
—Longfellow, 1863

Here's another sort of waiting problem, one that big-city dwellers probably face all the time. Imagine that you've just moved to a large city, and a friendly neighbor has told you about the local public transportation system. There are two independently operated bus lines, both of which stop in front of your apartment building. One stops every hour, on the hour. The other also stops once an hour, but not on the hour. That is, the first bus line arrives at ..., 6 a.m., 7 a.m., 8 a.m., etc., and the second bus line arrives at ..., $(6+x)$ a.m., $(7+x)$ a.m., $(8+x)$ a.m., etc., where x is a positive constant. Unfortunately, your neighbor (who walks to work and doesn't use either bus line) doesn't know the value of x. In the absence of any other information, then, let's assume that x is a particular value of random quantity uniformly distributed from 0 to 1. Our question is then easy to state: What is the average waiting time (until a bus arrives) that inexperienced visitors to the city can expect to wait if they arrive at the stop "whenever"?

This problem can be generalized in the obvious way, as follows: Let there be a total of n independently scheduled bus lines, all making hourly stops (only one is on the hour) in front of your apartment building ($n = 2$ is the above, original problem). What is the average waiting time until a bus of one of the bus lines arrives for a rider who arrives at the stop at random? Write a Monte Carlo simulation of this scenario and run it for $n = 1, 2, 3, 4$, and 5. To partially check your code, the theoretical answer for the $n = 1$ case is obviously one-half hour (30 minutes), while for the $n = 2$ case the answer is the not so obvious one-third hour (20 minutes). From your simulation results, can you guess the general answer for any value of n?

18. Waiting for Stoplights

Down these mean streets a man must go who is not himself mean, who is neither tarnished nor afraid....
—Raymond Chandler, "The Simple Art of Murder" (*Atlantic Monthly*, December 1944)

Yes, but even Philip Marlowe, Chandler's quintessential hard-boiled, tough-guy private eye of 1940s Los Angeles, waited (usually) for red traffic lights to turn green before crossing the street.

This waiting problem first appeared in the early 1980s. With reference to Figure P18.1, imagine a pedestrian starting out on a shopping trip in the Big City, m blocks east and n blocks north of her ultimate destination at $(1, 1)$. That is, she starts at $(m + 1, n + 1)$. (In the figure, $m = n = 2$, and she starts from $(3, 3)$). She slowly works her way down the streets, shopping and window gazing, letting the traffic lights at each intersection control which way she'll go next. In other words, as she reaches each intersection she always goes either west or south, with the direction selected being the one with the green light. This process will eventually bring her either to the horizontal boundary line $k = 1$ or to the vertical boundary line $j = 1$, at which point she will no longer have a choice on which way to go next. If she arrives at the horizontal

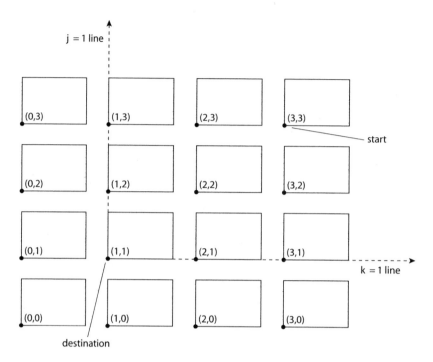

Figure P18.1. The big city.

line $k = 1$ she will thereafter have to always go west, i.e., to the left along that line, and if she arrives at the vertical line $j = 1$ she will thereafter have to always go south, downward along that line.

Until she hits one of the two boundary lines she never has to wait at an intersection, as there is always a green light in one of the two directions that moves her ever closer to (1, 1). Once on either of the boundary lines, however, she may well come to intersections at which the traffic light in the one direction she can move along happens, at that moment, to be red, and so she'll have to wait until the light turns green. And that's the question here: How many red lights, on average, will she have to wait for on her journey from $(m + 1, n + 1)$ to (1, 1)? Determine that average for each of the 1,001 cases of $0 \leq m = n \leq 1,000$ (notice that the average number of stops for the case of $m = 0$ is zero—use this as a check case) and plot it. Assume that when she reaches any intersection that the traffic lights there are as likely to be green (or red) in one direction as in the other direction.

19. Electing Emperors and Popes

You have chosen an ass.

- *The words of James Fournier, after his unanimous election in 1334 as Pope (Benedict XII). Whether he meant to be humble or ironic (or perhaps simply honest) is unclear.*

Imagine there is a group of N people who wish to elect one of themselves as leader of the group. Each member of the group casts a vote, in a sequence of secret ballots, until one of them receives at least M votes. Two historically interesting examples of this situation are the so-called Imperial Election problem[1] ($N = 7$ and $M = 4$, i.e., a simple majority is required to win) and the election of the pope (N is the membership of the College of Cardinals and M is the first integer equal to or greater than two-thirds of N). If each member of the group votes for one of the group's members (perhaps even for themselves) at random, what is the probability a leader will be chosen on any given ballot?

Let's now add a little twist by further assuming that rather than voting at random for any member of the group, all vote at random for one of their colleagues in a subset of N, of size $n \leq N$. All N people follow this restriction, including those in the subset. If $n = N$, we have the original problem. Obviously, $n > 0$ if we are to even have a vote.

And equally obvious is the answer for the $n = 1$ case: the probability of electing a leader on the first ballot is one. It is for $n \geq 2$ that the problem is of mathematical interest. This twist is motivated by a story told in Valérie Pirie's 1936 book *The Triple Crown*, a history of papal conclaves since 1458. In particular, the one of 1513 is said to have begun with such a divergence in support for a leader that, on the first ballot, all the voting cardinals independently decided to vote for one or another of those cardinals in a small subset of cardinals that were generally thought to have little (if any) support. They all did this, apparently, with each thinking they were the only ones doing so and thus would learn which way the wind was blowing. Much to the surprise (if not horror) of all, however, one of these unworthies received 13 votes, nearly enough to be elected pope (in 1513, $N = 25$ cardinals present and so $M = 17$). This would have been a disaster, as Pirie declared this almost-pope to be "the most worthless nonentity present."

Such disasters are said to actually have happened. In a famous eight-volume work by the dean of St. Paul's in London, Henry Hart Milman (*History of Latin Christianity*, 1861), for example, we find the following two passages; the first describes the conclave of 1334 (twenty four cardinals present) and the second the conclave of 1431 (fourteen cardinals present):

> In the play of votes, now become usual in the Conclave, all happened at once to throw away their suffrages on one for whom no single vote would have been deliberately given. To his own surprise, and to that of the College of Cardinals and of Christendom ... James Fournier found himself Pope [see the quote that opens this problem]. (vol. 7, p. 121)

> The contest lay between a Spaniard and a French Prelate. Neither would make concessions. Both parties threw away their suffrages on one whom none of the College desired or expected to succeed: their concurrent votes fell by chance on the Cardinal of Sienna [Pope Eugene IV]. (vol. 7, p. 538)

It has been mathematically argued, however, that these two events are most unlikely to have actually occurred as described by Milman.[2]

Estimating the probability of selecting the group leader by such random voting as a function of N, M, and n is duck soup for a computer, and your assignment here is to write a Monte Carlo simulation that accepts the values of N, M, and n as inputs. Use your code to estimate the probability for the case of $N = 7$ (the Imperial Election problem), and for $N = 25$ (the 1513 Pope problem[3]) run your code for the cases of $n = 2$, 3, and 4. Do these simulations for the two cases of (1) each person possibly voting for himself and (2) not being allowed to vote for himself.

References and Notes

1. The Imperial Election problem is motivated by the election of the emperor of the Holy Roman Empire by seven electors. The so-called prince-electors of the Holy Roman Empire (the predecessor to what we call Austria and Germany today) didn't, however, elect one of themselves. Some of the electors were actually archbishops and hardly candidates for the title of King of the Romans. The king became emperor only after being crowned in Rome by the pope. The Empire, abolished in 1806 after existing for 900 years, was famously described in 1756 by Voltaire: "This agglomeration which was called and which still calls itself the Holy Roman Empire was neither Holy, nor Roman, nor an Empire."

2. In his paper "An Unlikely History Story" (*The Mathematical Gazette*, December 1986, pp. 289–290), Professor Anthony Lo Bello of Allegheny College's math department explains: "If there are r cardinals, m of whom are not serious candidates, in how many different ways, we ask, can the cardinals vote among the m "dark horses," where the voting is random except for the fact that no one may vote for himself, and where we call two votes different if they are cast by different cardinals, even if they go to the same candidate? Since each of the m "dark horses" has $m - 1$ choices [can't vote for themselves], while each of the $r - m$ likely candidates has m choices, the number of allowed outcomes, which are all equiprobable, is

$$m^{r-m}(m-1)^m.$$

Of these outcomes, there are exactly $m(m-1)$ in which one of the 'long shots' receives $r - 1$ votes and some other one of them [of the long shots] gets the winner's vote [and who didn't, of course, vote for himself]. The probability that some "dark horse" ends up getting all the votes but his own is therefore

$$\frac{m(m-1)}{m^{r-m}(m-1)^m} = \frac{1}{m^{r-m-1}(m-1)^{m-1}}.$$

Professor Lo Bello then calculated the value of this expression for the two conclaves described by Milman. Under the assumption of $m = 5$ (and $r = 24$) for the 1334 conclave, we get the probability of the event described to be

$$\frac{1}{5^{24-5-1} \times (5-1)^{5-1}} = \frac{1}{5^{18} \times 4^4} = 1.02 \times 10^{-15}.$$

For the 1431, conclave we are told there were only two serious candidates, and so, with $r = 14$, we have $m = 12$. The probability of the event described by Milman is thus

$$\frac{1}{12^{14-12-1} \times (12-1)^{12-1}} = \frac{1}{12 \times 11^{11}} = 2.92 \times 10^{-13}.$$

These two probabilities are so small that, as Professor Lo Bello concluded, the two stories told by Milman "are thus fable[s] and can not be believed."

3. For the curious, the final result was the election of Giovanni de' Medici (Leo X), who served until 1521.

The Very Best

20. An Optimal Stopping Problem

"The very best" is the enemy of "good enough."
—an old saying among designers of military weapons systems

A good example of the above assertion is the cruise missile. A cruise missile is powered by a lightweight, inexpensive jet engine that, by its very nature, has a life expectancy after launch measured in a very small number of hours at most. The whole point of a cruise missile, after all, is to successfully vanish in a stupendous fireball once it arrives on target, and so why use a reliable, long-life, expensive engine? "Good enough" is all you need. When it comes to affairs of the heart, however, our practical engineering dictum loses all its force. When looking for a mate, who wants to settle for anyone less than perfect? But how do you find that special one, that unique soul-mate with whom to spend the next seventy years in wedded bliss? Well, as America's 50% divorce rate testifies, it ain't easy!

The generally accepted procedure is that of dating. That is, reducing the search for love to the level of searching for low-priced gasoline, one first gains experience by sampling for a while among potential mates, and then at some point a decision is made to buy. To start the casting of this situation into mathematical terms, let's suppose that the search for that special one obeys the following rules:

1. There is a known total population of potential special ones from which you can select *the* one;
2. You can date from the total population, one after the other, for as long as you like (or until you have dated the entire population);
3. After each date you have to decide if that person is *the* one— if not, then you date again if you haven't exhausted the population (if you have exhausted the population then you are stuck with the final date);
4. After you decide not to pursue the current date, you can never date that particular person again.

This situation was nicely discussed a few years ago in a popular, math-free book[1] by a physicist, who asked the following central question:

> The selection problem is obvious. You want the best spouse, but how can you maximize your [chances of finding him/her] under these rules? If you plunge too early in your dating career, there may be finer, undated fish in the sea, and you may go through life regretting a hasty marriage.... Yet if you wait too long the best may have slipped through your fingers, and then it is too late.

So, what should you do? Our physicist suggests the following general strategy where, just to have some specific numbers to talk about, he assumes that there is an initial population of one hundred potential special ones:

> [Y]ou shouldn't choose the first [date] who comes along—it would really be an amazing coincidence (a chance in a hundred) if the best of the lot showed up first. So it would make sense to use the first group of dates, say ten of them, as samplers ... *and then marry the next date who rates higher than any of [the samplers]*. That's a way of comparing them, and is not far from real life[2].... All you're doing is using the first ten dates to gain experience, and to rate the field. That's what dating is all about.

This strategy is only probabilistic, of course, and it doesn't guarantee that you will do well. In fact, as our physicist goes on to explain,

> There are two ways you can lose badly.... If the first ten just happen to be the worst of the lot—luck of the draw—and the next one just happens to be the eleventh from the bottom, you will end up with a pretty bad choice—not the worst, but pretty bad—without ever coming close to the best. You have picked the eleventh from the bottom because [he/she] is better than the first ten—that's your method—while the best is still out there.... The other way you can lose is the opposite: by pure chance the best may have actually been in the first ten, leading you to set an impossibly high standard.... You will then end up going through the remaining ninety candidates without ever seeing [his/her] equal, and will have to settle for the hundredth.... The hundredth will be, on average, just average.

The value of ten used above was just for purposes of explanation, as determining what should be the size of the sample lot is actually the core of the problem. And, in fact, there is a very pretty theoretical analysis that gives the value of that number (as a function of the total population of potential special ones that you start with), as well as the probability that using that number will result in your selecting the very best possible mate in the entire population. (For an initial population of 100, the optimal size of the sample lot is significantly greater than ten.) I'll show you the derivation of that formula in the solution, but for now let's assume that you don't know how to derive it (that's the whole point of this book, of course). This means, as you've no doubt already guessed, that you are to write a Monte Carlo simulation of this strategy. Your simulation should ask for the size of the initial population (call it n) and then, for all possible sizes of the sample lot (which can obviously vary from 0 up to $n - 1$), estimate the probability that you select the very best. The optimal value of the sample lot size is, of course, the value that maximizes the probability of selecting the very best person. Assume that each person has a unique ranking from 1 to n (the very best person has rank 1, and the worst person has rank n). The order in which you date people from the initial population is what is random in this problem. For a population of size n, there are

$n!$ possible orders (which quickly becomes enormous as n increases) and so for large n, your code will examine only a small fraction of all possible orders. In fact, write your simulation so that you will consider your quest for a mate a success as long as you select a person who is one of the top two (or three or four or ...), which includes, as a special case, that of selecting the very best. Your code will be very helpful in this generalization of the problem, as the pretty theoretical analysis I mentioned above holds only in the case of picking the very best. As some particular results to use in partially validating your simulation, for the case of $n = 11$ the theoretical results are:

Sample size	Probability of selecting the very best person
0	0.0909
1	0.2663
2	0.3507
3	0.3897
4	0.3984
5	0.3844
6	0.3522
7	0.3048
8	0.2444
9	0.1727
10	0.0909

Notice that the first and last rows make immediate sense. If the sample size is zero, that means you will select the first person you date, and that person is the very best person with probability $1/11 = 0.0909$. If the sample size is ten, then you will, by default, select the last person you date, and the probability that person is the best person is also $1/11 = 0.0909$. For $n = 11$, we see from the table that the sample lot size should be four; i.e., you should date (and reject) the first four people you meet while remembering the best of those four, and then propose to the first person thereafter who is better (has a smaller ranking number) than the best you saw in the first four. This strategy will result in you proposing to the very best person with a probability of 0.3984 (which is, I think, surprisingly high).

Use your code to answer the following questions:

1. For $n = 11$, what are the optimal sample sizes (and what are the associated probabilities of success) if you are happy if you select a person from the top two, three, four, and five?
2. For $n = 50$, what is the optimal sample size if you insist on being happy only if you select the very best person?
3. For $n = 50$, repeat the simulations of (1).

References and Notes

1. H. W. Lewis, *Why Flip a Coin? The Art and Science of Good Decisions*, (New York: John Wiley & Sons, 1997, pp. 4–11) ("The Dating Game").

2. An amusing illustration of the dating game from the history of science, can be found in the second marriage of the German astronomer Johannes Kepler (1571–1630). (Kepler is famous for his formulation of the three planetary laws of motion from direct observational data. The laws were later shown in Isaac Newton's *Principia* [1686] to be mathematical consequences of the fact that gravity is an inverse-square, central force.) After the death of his first wife—an unhappy, arranged union—Kepler decided to take matters into his own hands and to choose a wife by rational means: he interviewed eleven women (this is why I used $n = 11$ as an example at the end of the above discussion) before arriving at his decision! The rules of Kepler's dating were not the same as the rules we are using here, but still, the spirit of the dating game is clearly centuries old. You can find all the fascinating details in the biography of Kepler by Arthur Koestler, *The Watershed* (New York: Anchor Books, 1960); in particular, see Chapter 10, "Computing a Bride" (pp. 227–234).

21. Chain Reactions, Branching Processes, and Baby Boys

The trees in the street are old trees
used to living with people,
Family trees that remember your
grandfather's name.
—Stephen Vincent Benét, *John Brown's Body*

Initially I toyed with the idea of making the previous problem (with its suggestive title) the last problem in the book. But with this problem we'll end with what comes after (usually) all the dating—babies! But before I formulate the central problem to be posed here, let me start by discussing what might seem to be a totally off-the-wall and unrelated topic—atomic bombs! You'll soon see the connection.

One can't read newspapers or watch television and not be aware that the worldwide fear of nuclear weapons is as strong today as it was in the 1950s, at the height of the cold war between the United States and the old Soviet Union. Today both the Americans and the Russians still have their thousands of nuclear weapons, as do, in lesser numbers, the French, British, Israeli, Chinese, Pakistani, and Indian governments. Seemingly on the verge of joining this elite group are (as I write) North Korea and Iran. Modern nuclear weapons are vastly more sophisticated than were the first atomic detonation in the Alamagordo, New Mexico, desert and the two atomic bombs dropped by America on Japan in 1945, but the underlying mechanism of atomic fission is still of interest today. Atomic fission bombs are measured in tens or hundreds of kilotons (of TNT) of destructive energy, while the newer thermonuclear fusion bombs are measured in megatons of explosive energy. It still takes a fission process to achieve the enormous temperatures (at least 70 million degrees!) and pressures (that at the center of the Earth!) required to initiate or trigger a fusion process, however, and so fission is of continuing interest.

The question of how fission weapons work seems, therefore, to be one appropriate to any discussion of technically significant practical problems and, as I discuss next, probability plays a central role. Imagine a blob of what is called a fissionable material (an isotope of uranium, U-235, is a well-known example; another is plutonium-239), by which is meant stuff whose atoms are borderline unstable. (The numbers 235 and 239 refer to the so-called atomic weight of the uranium and plutonium atoms, respectively, a concept we don't have to pursue here.) That is, with the proper external stimulus any of those atoms can be made to come apart, or split, into what are called fission fragments, along with the release of some energy. Each such fission produces only a very tiny amount of energy, but with 2.5×10^{24} atoms in each kilogram of U-235 or Pu-239, it doesn't take a lot of either to make a really big hole in the ground. The first atomic bomb dropped on Japan (Hiroshima), for example, fissioned less than one kilogram of U-235 (out of a total fissionable mass of sixty kilograms, a mass with the volume of a grapefruit) but produced an explosion equivalent to fifteen thousand tons of TNT. Similarly for plutonium, as the second Japan bomb (Nagasaki) fissioned just a bit more than one kilogram of Pu-239 (out of a total fissionable mass of 6.2 kilograms), with the result being a 23-kiloton explosion. (Totally fissioning one kilogram of either U-235 or Pu-239 produces what was, and may still be, called the nominal atomic bomb, equivalent to about twenty kilotons of TNT.)

The proper external stimulus is an energetic, electrically neutral particle (a neutron) that plows unimpeded through an atom's outer electron cloud and into that atom's nucleus, and so splits that nucleus.

Among the resulting fission fragments are two or three more neutrons that are now able to go on to potentially smash into other nearby nuclei, thereby creating even more neutrons. And so we have the famous chain reaction. If the reaction goes forward in such a way that the cascade of neutrons grows ever larger, then the total release of energy can be stupendous. It is a distinguishing characteristic of chain reactions that they develop with fantastic speed, with the time lapse between a neutron being released as a fission fragment until it in turn causes the next level of fission being something like 10 nanoseconds (10×10^{-9} seconds). Therefore, in just one-half microsecond (0.5×10^{-6} second), there can be fifty so-called chain reaction generations. Now, if each fission produces an average of 2.5 neutrons (e.g., half the fissions produce two neutrons and the other half result in three neutrons), each of which goes on to cause another fission, then, starting with one neutron (the zero-th generation), we have the exponentially growing sequence 1, 2.5, 2.5^2, 2.5^3, ... After the nth generation we will have generated a total of $2.5 + 2.5^2 + 2.5^3 + \cdots + 2.5^n$ neutrons, i.e., a geometric series easily summed to give a total of $2/3 \times (2.5)^{n+1}$ neutrons. If we set this equal to 2.5×10^{24} (that is, if we imagine we have totally fissioned a one-kilogram blob of either U-235 or Pu-239) we find that $n = 61$ (which, from start to finish of the chain reaction, takes just 0.61 microseconds). The release of a stupendous amount of energy in less than a microsecond is unique to atomic explosions. By comparison, ordinary chemical explosions involve vastly less energy released over time intervals measured in milliseconds; i.e., chemical explosions are both less energetic and a thousand times (or more) slower than atomic explosions. Notice, too, that in an exponential chain reaction explosion most of the energy is released in just the final few generations.

One problem with all the above, however, is that not all the neutrons in the chain reaction cascade of neutrons necessarily result in subsequent fissions. Not even the very first neutron has that guarantee, and if it fails to cause a fission, then of course there is no neutron cascade at all, and the bomb is a dud from the start.[2] It was, therefore, an important problem in the American atomic bomb program (code-named the Manhattan Project) to study neutron chain reaction cascades with this important realistic complication taken into account. The problem was mathematically formulated in 1944 by the Manhattan Project mathematician Stanislaw Ulam as follows: Suppose p_i, $i = 0, 1, 2, 3, \ldots$, are the probabilities a neutron will result in a fission that produces i new neutrons. In particular, then, p_0 is the probability a neutron does not result in a fission (the only way to get zero neutrons from an initial neutron is to not have a fission). Ulam's question, then, was that of calculating the probability distribution of the number of neutrons produced in the nth generation of a chain reaction cascade, with a single neutron as the lone occupant of the 0th generation.

Ulam thought his neutron chain reaction question was new with him, but later learned (after he and his Los Alamos colleague David Hawkins solved it[1]) that it was actually a hundred years old! In 1874 an analysis of the extinction probability of a family name in the British peerage had been done by the English mathematicians Sir Francis Galton (1822–1911) and Henry Watson (1827–1903), and they in turn had been anticipated by the French mathematician Irénée-Jules Bienaymé (1796–1878), who had written on a similar problem in 1845. The family name extinction problem is simply this: as Hawkins writes, "Assume that each male Jones produces i male offspring with probability p_i (and that this probability does not vary from one generation to another). What then is the probability that a given Jones has k males in the nth generation of his descendents?" The probability for $k = 0$ is the probability of the family name of Jones vanishing at the nth generation. Just substitute "neutron" for "male offspring" and you have the nineteenth-century family name problem transformed into Ulam's twentieth-century neutron chain reaction problem. Hawkins's explanation of the family name extinction problem is, however, incomplete on one crucial issue: we are interested only in those males in each generation who trace their ancestors back to the initial lone male strictly through their fathers. That is, there can be male descendents who are the result of daughters in the family tree, but of course in that case the original family name does not continue.

There is an elegant theoretical solution to what is now called a problem in stochastic branching processes, and I'll tell you a little bit about it in the simulation solution, where I'll use it to partially verify my Monte Carlo code, but for now concentrate on developing that code yourself. When you write your code, use the result of a study done in the 1930s by the American mathematician Alfred Lotka (1880–1949), who showed that, at least in America, the p_i probabilities are closely approximated by $p_0 = 0.4825$ and $p_i = (0.2126)(0.5893)^{i-1}$ for $i \geq 1$. Specifically, what does your code estimate for the probability of there being two male descendents in the second generation? Of there being four male descendents in the second generation? Of there being six male descendents in the third generation? Lotka's formula assumes that a man can have any number of sons, which is clearly not reasonable. In your simulation, therefore, assume that no male ever has more than seven sons. This will allow the (unlikely) possibility of there being, in the third generation, a seventh son of a seventh son of a seventh son—that possibility has little mathematical significance, but it does appeal to my mystical side!

References and Notes

1. David Hawkins, "The Spirit of Play: A Memoir for Stan Ulam" (*Los Alamos Science* [Special Issue], 1987, pp. 39–51).

2. For this reason, all practical atomic fission bomb designs include a built-in neutron generator that floods the bomb core with neutrons at the instant before the desired instant of detonation, a flood that ensures there will indeed be a detonation.

The Solutions

1. The Clumsy Dishwasher Problem

We can calculate the theoretical probability as follows. There are $4\binom{5}{4} + \binom{5}{5}$ ways to assign any four, or all five, of the broken dishes to "clumsy." The factor of 4 in the term $4\binom{5}{4}$ is there because, after we have "clumsy" breaking four dishes, the remaining fifth broken dish can be assigned to any of the other four dishwashers. There are a total of 5^5 different ways to assign the broken dishes among all five dishwashers. So, the answer to the question of the probability of "clumsy" breaking at least four of the five broken dishes *at random* is

$$\frac{4\binom{5}{4} + \binom{5}{5}}{5^5} = \frac{20+1}{3,125} = \frac{21}{3,125} = 0.00672.$$

The code dish.m simulates the problem, assuming the broken dishes occur at random, where the variable brokendishes is the number of dishes broken by "clumsy." If the value of the variable clumsy is the total number of times "clumsy" breaks four or more dishes (that is, clumsy is the total number of simulations in which brokendishes > 3), then, with each such simulation, clumsy is incremented by one. When dish.m was run, line 14 produced an estimate for the probability that "clumsy" breaks at least four of the five broken dishes, due strictly to random chance, as 0.00676 (when run several times for just 10,000 simulations, the estimates varied from 0.0056 to 0.0083). Theory and experiment

are in pretty good agreement, and in my opinion, this probability is
sufficiently small that a reasonable person could reasonably conclude,
despite his denials, that "clumsy" really is clumsy! With some non-zero
probability (see above), of course, that "reasonable" conclusion would
actually be incorrect.

---

dish.m

```
01    clumsy = 0;
02    for k = 1:1000000
03        brokendishes = 0;
04        for j = 1:5
05            r = rand;
06            if r < 0.2
07                brokendishes = brokendishes + 1;
08            end
09        end
10        if brokendishes > 3
11            clumsy = clumsy + 1;
12        end
13    end
14    clumsy/1000000
```

2. Will Lil and Bill Meet at the Malt Shop?

Let's denote the arrival times of Lil and Bill by L and B, respectively, where L and B are taken as independent and uniformly distributed random variables over the interval 0 to 30 (0 means arriving at 3:30 and 30 means arriving at 4 o'clock). Obviously, if $L < B$, then Lil arrives first, and otherwise ($L > B$) Bill arrives first. So, we can now immediately write

$$if \ L < B, \text{ then Lil and Bill will meet } if \ B - L < 5$$

and

$$if \ L > B \text{ then Lil and Bill will meet } if \ L - B < 7.$$

Figure S2.1 shows the geometric interpretation of these two inequalities. The 30×30 square and its interior points (called the *sample space* of the problem by mathematicians) represents the infinity of all possible pairs of arrival times for Lil and Bill—its diagonal line $L = B$ is the separation between the two possibilities of Lil arriving first and Bill arriving first (the third possibility, Lil and Bill arriving simultaneously, has zero probability, and is represented by the collection of all the sample space points *on* the diagonal line).

With a little manipulation we can write these two inequalities in the following alternative way:

$$\text{Lil and Bill will meet } if \ B < L < B + 7$$

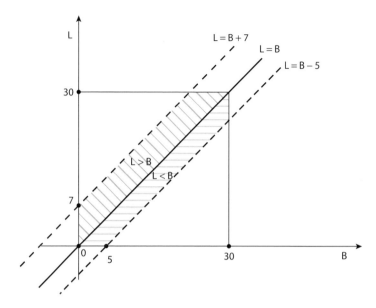

Figure S2.1. Geometric interpretation of Lil and Bill's meeting inequalities.

and

Lil and Bill will meet *if* $B - 5 < L < B$.

The area of sample space corresponding to the first of these two double inequalities is the shaded area above the sample space diagonal, and the area of sample space corresponding to the second double inequality is the shaded area below the sample space diagonal. According to the fundamental idea underlying geometric probability, we can write

$$\text{probability Lil and Bill meet} = \frac{\text{shaded area}}{\text{total area of sample space}}$$

$$= \frac{(\text{total area above diagonal} - \text{unshaded area above diagonal})}{\text{total area of sample space}}$$

$$+ \frac{(\text{total area below diagonal} - \text{unshaded area below diagonal})}{\text{total area of sample space}}$$

$$= \frac{\left(450 - \frac{1}{2} \times 23 \times 23\right) + \left(450 - \frac{1}{2} \times 25 \times 25\right)}{900}$$

$$= \frac{900 - \frac{1}{2}(23^2 + 25^2)}{900} = 1 - \frac{529 + 625}{2 \times 900} = 1 - \frac{1,154}{1,800}$$

$$= 0.358889.$$

The code malt.m is a Monte Carlo simulation of the problem, which directly implements our two double inequalities for a total of one million potential meetings. When run, line 11 produced an estimate for the probability of Lil and Bill meeting of 0.359205, which is in good agreement with the theoretical calculation.

---

malt.m

```
01      meetings = 0;
02      for loop = 1:1000000
03          L = 30*rand;
04          B = 30*rand;
05          if B<L&L<B + 7
06              meetings = meetings + 1;
07          elseif L<B&L>B − 5
08              meetings = meetings + 1;
09          end
10      end
11      meetings/1000000
```

If Bill reduces his waiting time to five minutes, then the probability of meeting becomes

$$1 - \frac{\frac{1}{2}(25^2 + 25^2)}{900} = 1 - \frac{1,250}{1,800} = 0.305556,$$

and if Lil increases her waiting time to seven minutes, then the probability of meeting becomes

$$1 - \frac{\frac{1}{2}(23^2 + 23^2)}{900} = 1 - \frac{1,058}{1,800} = 0.412222.$$

If we change line 05 to if B<L&L<B+5, then malt.m gives the estimate 0.305754 for the meeting probability in the first case, and if we change line 07 to elseif L<B&L>B−7, then malt.m gives the estimate 0.41199 for the meeting probability in the second case. These estimates are seen to be in pretty good agreement with the theoretical calculations.

3. *A Parallel Parking Question*

This problem originally appeared in 1978, as a challenge question. As stated then,[1]

> Four birds land at random positions on a finite length of wire. Upon landing, each bird looks at its nearest neighbor.
>
> a. What is the probability that a bird picked at random is looking at another bird that is looking at it?
> b. What is the probability if there are n birds, with $n > 4$?

The solution appeared the next year.[2] (I have converted the birds into cars and the wire into a car lot to, I hope, make the problem more "relevant".)

A Monte Carlo simulation for this problem is provided by the code car.m, which uses the two row vectors pos and nn. The vector pos is defined to be such that pos(j) gives the position of the jth car ($1 \leq j \leq n$), where $0 \leq \text{pos}(j) \leq 1$; we'll say that the jth car has the *number-name j*. It is assumed that pos(j) monotonically increases as j increases, and you'll see how that property is established in just a bit. The vector nn is the *nearest neighbor* vector, defined to be such that nn(j) is the number-name of the nearest neighbor car of the jth car, e.g., nn(1) = 2 and nn(n) = $n - 1$, always, for any integer values of $n \geq 2$. The code car.m works as follows.

Line 01 initializes the variable totalmn to zero (this variable's role will be explained later), and line 02 sets the value of the variable n, which, of course, is the number of cars. Line 03 is the start of the outermost loop, the loop that controls a total of one million simulations of n

cars parked in the lot. Line 04 is the start of an individual simulation; lines 04 through 06 assign, at random, the parking positions of the n cars. Those positions are not, however, necessarily such that pos(1) < pos(2)< $\cdots$ < pos(n), and so lines 07 through 20 sort the position vector elements to ensure that a monotonic condition is indeed satisfied.

There are numerous sorting algorithms available, but I have used one commonly called, by computer scientists, the *bubble-sort* algorithm.[3] It works by comparing each value in the pos row vector with the value next to it (to the "right") and interchanges the two values, if necessary, to achieve a local increase in values. The algorithm continually loops through pos, repeating this process until it makes a complete pass through the pos vector without any swapping occurring (that condition is determined by the value of the variable check— if check = 0 at line 08, then no swapping was done on the last pass through pos), which means pos is now guaranteed to be in monotonic increasing order from left to right. This ordering has occurred because the larger values in pos have "bubbled" their way to the right, while the smaller values in pos have drifted to the left. The bubble-sort algorithm is notoriously inefficient, but it is very easy to understand—for a programmer in his mid-sixties (your author), that's a very big plus!— and for small values of n (say, n < 100), the computational inefficiency is of no real practical concern. On my quite ordinary computer, for example, a million simulations with $n = 30$ takes only about seventy-five seconds. Lines 21 through 29 then use the sorted pos vector to create, in the obvious way, the row vector nn.

Lines 30 through 42 are the heart of car.m, using nn to determine the number of pairs of cars that are mutual neighbors. The total number of cars that are members of the mutual pair set is, of course, twice the number of pairs. At the completion of a simulation, the variable mn will be the number of *mutual neighbor* pairs; line 30 initializes mn to zero. Lines 31 through 33 treat the special starting case of determining if the first two cars, located at nn(1) and nn(2), are a mutual neighbor pair. Thereafter the rest of the pos vector is examined by lines 34 through 42 to find all the remaining mutual pairs. Starting with j=2, the code checks to see if nn(j) and nn(j+1) are the locations of a mutual neighbor pair—if so, then mn is updated and j is incremented by two. If not, then j is simply incremented by one and the checking process repeated until the entire pos vector has been examined. To see how the results look,

car.m

```
01   totalmn = 0;
02   n = input('What is n?');
03   for loop = 1:1000000
04       for k = 1:n
05           pos(k) = rand;
06       end
07       check = 1;
08       while check == 1
09           check = 0;
10           i = 1;
11           while i < n
12               if pos(i) > pos(i + 1)
13                   temp = pos(i + 1);
14                   pos(i + 1) = pos(i);
15                   pos(i) = temp;
16                   check = 1;
17               end
18               i = i + 1;
19           end
20       end
21       nn(1) = 2;
22       nn(n) = n − 1;
23       for j = 2:n − 1
24           if pos(j) − pos(j − 1)<pos(j + 1) − pos(j)
25               nn(j) = j − 1;
26           else
27               nn(j) = j + 1;
28           end
29       end
30       mn = 0;
31       if nn(2) == 1
32           mn = mn + 1;
33       end
34       j = 2;
35       while j<n
36           if nn(j) == j + 1&nn(j + 1) == j
```

(continued)

(continued)

```
37                          mn = mn + 1;
38                           j = j + 2;
39              else
40                           j = j + 1;
41              end
42         end
43         totalmn = totalmn + mn;
44    end
45    2*totalmn/(1000000*n)
```

when one run of car.m was looked at in detail, for the $n = 7$ case, the pos and nn vectors were:

pos = [0.0210   0.2270   0.2274   0.3853   0.6461   0.6896   0.7971]
nn = [2   3   2   3   6   5   6].

You should be sure you understand why these two vectors are consistent. Once mn is calculated at the end of each simulation, totalmn is updated in line 43. At the completion of one million simulations, line 45 gives us the probability of a randomly selected car being one of a nearest neighbor pair (2*totalmn is the total number of such cars out of a total of 1000000*n cars).

When car.m was run, it produced the following results:

n	Probability
3	0.66667
4	0.66628
5	0.66697
6	0.6665
7	0.66674
8	0.66671
9	0.66691
10	0.66675
11	0.66684
12	0.66667
20	0.66666
30	0.66662

One cannot look at these numbers without immediately being struck by the nearly constant value of the probability, even as n changes by a lot. That is, car.m strongly suggests that this probability is always, for any $n \geq 3$, equal to $\frac{2}{3}$. This is, I think, not at all intuitively obvious! A clever proof[2] of this suggestion appeared in 1979, and two years later a more direct proof was published[4] that had the added virtue of being extendable, in an obvious way, to the cases of two-dimensional and three-dimensional parking lots (the last case is, of course, simply an ordinary multilevel parking garage). There it was shown that, in the limit as $n \to \infty$, the probabilities for a randomly selected car being one of a nearest neighbor pair are:

$$\text{for two dimensions,} \qquad \frac{6\pi}{8\pi + 3\sqrt{3}} = 0.621505$$

and

$$\text{for three dimensions,} \qquad \frac{16}{27} = 0.592592.$$

References and Notes

1. Daniel P. Shine, "Birds on a Wire" (*Journal of Recreational Mathematics*, 10[3], 1978, p. 211).

2. John Herbert, "Birds on a Wire (Solution)" (*Journal of Recreational Mathematics*, 11[3], 1979, pp. 227–228).

3. I could have simply used MATLAB's built-in sorting command sort: if x is a row vector of n elements, then y = sort(x) creates the vector y with the same n elements arranged in ascending order from left to right; that is, the element values increase with increasing value of the vector index. If your favorite language doesn't have a similar handy command, then you'll have to write the sorting algorithm yourself. In the solution to Problem 17, however, where a sorting procedure also occurs, I reversed myself and elected to use sort. As I neared the end of my writing I became ever more willing to let MATLAB do the heavy lifting!

4. Charles Kluepfel, "Birds on a Wire, Cows in the Field, and Stars in the Heavens" (*Journal of Recreational Mathematics*, 13[4], 1980–81, pp. 241–245).

4. *A Curious Coin-Flipping Game*

The code broke.m simulates this problem. But first, before getting into its details, let me give you the theoretical answer[1] for the fair coins case. It is amazingly simple: the average number of tosses until one of the men is ruined is $\frac{4lmn}{3(l+m+n-2)}$. So, if $l=m=n=1$, the answer is $\frac{4(1)(1)(1)}{3(1+1+1-2)}=\frac{4}{3}$ as given in the problem statement. And for $l=1$, $m=2$, and $n=3$, the answer is $\frac{4(1)(2)(3)}{3(1+2+3-2)}=\frac{24}{12}=2$. If $l=2$, $m=3$, and $n=4$, the answer is $\frac{4(2)(3)(4)}{3(2+3+4-2)}=\frac{32}{7}=4.5714$; if $l=m=n=3$, the answer is $\frac{4(3)(3)(3)}{3(3+3+3-2)}=\frac{36}{7}=5.1428$; and if $l=4$, $m=7$, and $n=9$, the answer is $\frac{4(4)(7)(9)}{3(4+7+9-2)}=\frac{1,008}{54}=18.6667$. For $p=0.4$, we have no theory to guide us (as I'll elaborate on a bit at the end, this is not strictly true), and the simulation will be essential.

Here's how broke.m works, where it is assumed that the values of l, m, n, and p have already been input (see line 02 of car.m in the solution for Problem 3 for how MATLAB does that). The code simulates 100,000 coin-flipping sequences, and the value of the variable totalflips is the total number of simultaneous flips that the three men execute during the 100,000 sequences. Line 01 initializes totalflips to zero. The loop defined by the for/end statements of lines 02 and 44 cycle broke.m through the 100,000 sequences. During each *individual* sequence, the value of the variable sequenceflips is the number of simultaneous flips the three men execute, and line 03 initializes sequenceflips to zero. The three-element vector man is defined to be such that the value of man(j) is the current number of coins possessed by the jth man, $1 \le j \le 3$. Lines 04, 05, and 06 initialize the elements of man to their starting values of l, m, and n. The playing of a sequence then continues until the first time one of the elements of man is driven to zero; the while/end loop of lines 07 and 40 control this play. Lines 08, 09, and 10 represent the outcomes of the current simultaneous triple flip by the three men: the three-element vector flip is first randomly loaded, and then lines 11 through 17 map the elements of flip into either 0 or 1. Specifically, flip(j) is mapped into 1 with probability p, and into 0 with probability $1 - p$. To see why this is done, consider the following.

There are eight possible mappings for the elements of the flip vector, as shown in the three leftmost columns of the following table. The fourth column shows the sum of the flip vector elements (a calculation

---

broke.m
```
01    totalflips = 0;
02    for sequences = 1:100000
03        sequenceflips = 0;
04        man(1) = l;
05        man(2) = m;
06        man(3) = n;
07        while man(1)>0&man(2)>0&man(3)>0
08            flip(1) = rand;
09            flip(2) = rand;
10            flip(3) = rand;
11            for j = 1:3
12                if flip(j)<p
13                    flip(j) = 1;
14                else
15                    flip(j) = 0;
16                end
17            end
18            test = sum(flip);
19            if test == 1|test == 2
20                if test == 1
```

(continued)

(continued)

```
21                          for j = 1:3
22                              if flip(j) == 0
23                                  flip(j) = -1;
24                              else
25                                  flip(j) = 2;
26                              end
27                          end
28                      else
29                          for j = 1:3
30                              if flip(j) == 0
31                                  flip(j) = 2;
32                              else
33                                  flip(j) = -1;
34                              end
35                          end
36                      end
37                      for j = 1:3
38                          man(j) = man(j) + flip(j);
39                      end
40                  end
41                  sequenceflips = sequenceflips + 1;
42              end
43              totalflips = totalflips + sequenceflips;
44      end
45      totalflips/100000
```

performed in line 18, using the handy MATLAB command sum, which replaces the obvious alternative of a for/end loop), and the rightmost column shows the name of the winner (if any) on the current flip. Remember, the odd man out wins.

Notice that when test = 1, the winner is the jth man, where flip(j) = 1, and when test = 2, the winner is the jth man, where flip(j) = 0. This observation suggests a second mapping of the elements in the flip vector that we'll then be able to use to update the man vector. Specifically, the if in line 19 determines whether test is 1 or 2, or either 0

flip(1)	flip(2)	flip(3)	$test = \sum_{j=1}^{3} flip(j)$	Winner
0	0	0	0	None
0	0	1	1	Man 3
0	1	0	1	Man 2
0	1	1	2	Man 1
1	0	0	1	Man 1
1	0	1	2	Man 2
1	1	0	2	Man 3
1	1	1	3	None

or 3 (these last two possibilities are the cases of no winner, which results in broke.m skipping over the man vector update). If test is 1, then lines 21 through 27 map flip(j) = 0 into flip(j) = −1 and flip(j) = 1 into flip(j) = 2. Notice that this second mapping of flip has been cleverly designed to represent the winnings of the jth man, and so man(j) can be updated by simply adding flip(j) to man(j)—see line 38. Similarly, if test is 2 then lines 29 through 35 map flip(j) = 0 into flip(j) = 2, and flip(j) = 1 into flip(j) = −1, which again agrees with the winnings of the jth man. Lines 37, 38, and 39 then update the man vector. Line 41 increments the number of tosses completed so far in the current sequence and, once that sequence is done (as determined by the while condition in line 07), line 43 updates totalflips. When 10,000 sequences are completed, line 45 gives us the average number of tosses until one of the men is ruined.

How well does broke.m perform? The following table gives us an idea:

l	m	n	Theoretial answer	broke.m's estimate ($p = \frac{1}{2}$)	Estimate if $p = 0.4$
1	1	1	1.3333	1.335	1.3877
1	2	3	2	2.0022	2.0814
2	3	4	4.5714	4.5779	4.7721
3	3	3	5.1428	5.161	5.36
4	7	9	18.6667	18.8065	19.4875

Notice that *decreasing* p from 0.5 to 0.4 *increases* the average number of tosses until one of the men is ruined. Can you think of a reason for this, i.e., of an argument that one could use to arrive at this conclusion without having to actually run broke.m? Think about this for a while, as I show you a way we can analytically treat this problem, for unfair coins, if the total number of coins is small. In particular, I'll show you how to calculate the answer for any p in the case $l = m = n = 1$. The method can be extended to any set of values for l, m, and n, but the computations involved increase exponentially as $l + m + n$ increases, and the difficulties soon overwhelm anyone's wish to know the answer!

Let $E(l, m, n)$ denote the average number of tosses until one of the men is ruined. The probability that all three coins show the same side on a toss—and so there is no winner, and thus no coins change hands—is $p^3 + (1 - p)^3 = 1 - 3p + 3p^2$. In this event, $E(l, m, n)$ tosses are still expected in addition to the one toss that was just done. The probability the first man wins is $p(1 - p)^2 + (1 - p)p^2 = p(1 - p)$. In this event, $E(l + 2, m - 1, n - 1)$ tosses are still expected. By symmetry, we can say the same thing for the second and the third man. Thus,

$$E(l, m, n) = [1 + E(l, m, n)](1 - 3p + 3p^2)$$
$$+ [1 + E(l + 2, m - 1, n - 1)]p(1 - p)$$
$$+ [1 + E(l - 1, m + 2, n - 1)]p(1 - p)$$
$$+ [1 + E(l - 1, m - 1, n + 2)]p(1 - p).$$

Now, $E(l, m, n) = 0$ if any of l, m, or n is zero. So immediately we have

$$E(1, 1, 1) = [1 + E(1, 1, 1)](1 - 3p + 3p^2) + 3p(1 - p).$$

A little easy algebra gives us

$$E(1, 1, 1)[1 - (1 - 3p + 3p^2)] = E(1, 1, 1)(3p - 3p^2)$$
$$= 1 - 3p + 3p^2 + 3p(1 - p) = 1$$

or

$$E(1, 1, 1) = \frac{1}{3p - 3p^2} = \frac{1}{3p(1 - p)}.$$

If $p = \frac{1}{2}$ then $E(1, 1, 1) = \frac{1}{3 \times \frac{1}{4}} = \frac{4}{3}$, just as our general formula for fair coins does. But, now we know the theoretical answer for $E(1, 1, 1)$ for any p; for $p = 0.4$ we have $E(1, 1, 1) = \frac{1}{3(0.4)(0.6)} = 1.3889$, which agrees well with the estimate in the table provided by broke.m.

Our formula for $E(1, 1, 1)$ shows that, for $p = \frac{1}{2}$, $E(1, 1, 1)$ is the smallest it can be. As p deviates more and more from $\frac{1}{2}$ in either direction, $E(1, 1, 1)$ increases, just as reported by broke.m. The reason for this is that as $p \to 0$ or as $p \to 1$, it becomes more and more likely that all three coins will show the same face on a toss. Since such an outcome results in no coins changing hands, this raises the number of tosses we expect to see before one of the men is ruined.

References and Notes

1. R. C. Read, "A Type of 'Gambler's Ruin' Problem" (*American Mathematical Monthly*, February 1966, pp. 177–179).

5. *The Gamow-Stern Elevator Puzzle*

Rather than simulate the two-elevator and the three-elevator problems separately, the code gs.m simulates the general n-elevator problem for any integer $n \geq 1$. We imagine that the height of the building is scaled so that floor 1 (the bottom stop of an elevator) is at height 0 and floor 7 (the highest stop of an elevator) is at height 1. Thus, Gamow's elevator stop on the second floor is at height $G = \frac{1}{6}$. We further imagine, as described by Knuth, that all the elevators are, independently, at the moment that Gamow requests elevator service, at random heights, as well as independently moving either up or down with equal probability. The heart of gs.m is the n-by-4 array called elevator, in which the jth row describes the state of the jth elevator. Specifically,

> elevator(j,1) = direction (up or down) the jth elevator is moving when Gamow requests service;
>
> elevator(j,2) = height of the jth elevator when Gamow requests service;
>
> elevator(j,3) = distance the jth elevator has to travel to reach Gamow's stop;
>
> elevator(j,4) = direction (up or down) the jth elevator is moving when that elevator reaches Gamow's stop.

In gs.m "down" is coded as 0 and "up" is coded as 1.

We can now deduce the following four rules, for each elevator.

- if the height $< G$ and the initial elevator direction is down, then the travel distance to Gamow's stop is $G +$ height, and the arrival direction is up;
- if the height $< G$ and the initial elevator direction is up, then the travel distance to Gamow's stop is $G -$ height, and the arrival direction is up;
- if the height $> G$ and the initial elevator direction is down, then the travel distance to Gamow's stop is height $- G$, and the arrival direction is down;
- if the height $> G$ and the initial elevator direction is up, then the travel distance to Gamow's stop is $(1 -$ height$) + \frac{5}{6}$ $= \frac{11}{6} -$ height, and the arrival direction is down.

To answer the central question of this problem—what is the probability that the first elevator arriving at Gamow's stop is going in the "wrong" direction, i.e., down?—all we need do is to calculate which of the n elevators has the *least travel distance* to traverse, and to then observe in which direction that elevator is moving when it arrives at Gamow's stop. With these preliminary observations out of the way, you should now be able to follow the logic of gs.m. Lines 01, 02, 03, and 04 are initialization commands: setting $G = \frac{1}{6}$ (a constant used so often in the simulation that it is computationally far better to calculate it just once rather than to continually repeat the division); inputting the number of elevators, n; setting the variable totalgoingdown to zero, which, when gs.m ends, will be the total number of elevators that are the first to arrive at Gamow's floor while moving in the down direction; and defining the n-by-4 array elevator. Lines 05 through 41 are the for/end loop that runs gs.m through one million requests by Gamow for elevator service. An individual request is simulated by lines 06 through 29.

For each of the n elevators, lines 07 through 12 assign the up or the down direction, with equal probability, at the moment Gamow requests service. This establishes all the values of elevator(j,1), $1 \leq j \leq n$. Line 13 places each elevator at a random height, which establishes all the values of elevator(j,2). Then, lines 14 through 28 implement the four rules we deduced above, which establishes all the values of elevator(j,3) and elevator(j,4). Lines 30 through 37 search through all the values of elevator(k,3) to find that value k = index for which the elevator travel distance to Gamow's stop is minimum. Line 38 determines if this first-to-arrive elevator is traveling in the down direction, and if it is, the variable totalgoingdown is incremented. Line 42 gives us an estimate of the probability we are interested in once the one million service requests have been simulated.

Now, before I tell you how well gs.m performs, let me give you the theoretical answers. Knuth showed, in two different (both extremely clever) ways that for the original Gamow-Stern problem the first-to-arrive elevator at Gamow's stop is going down with probability $\frac{1}{2} + \frac{1}{2}\left(\frac{2}{3}\right)^n$. (Knuth actually solved the more general problem of a building with any number of floors, with Gamow's floor as any one of the building's floors.) For $n = 1$ we see this probability is, as Gamow and Stern argued, equal to $\frac{1}{2} + \frac{2}{6} = \frac{5}{6} = 0.833333$. For $n = 2$, Knuth's answer is $\frac{1}{2} + \frac{4}{18} = \frac{13}{18} = 0.722222$, as given in the problem statement. And, for $n = 3$, Knuth's formula gives the probability as $\frac{1}{2} + \frac{8}{54} = \frac{35}{54} = 0.648148$.

When gs.m was run, the estimates it produced for the $n = 1, 2$, and 3 cases were 0.833579, 0.722913, and 0.647859, respectively. This is, I think, in good agreement with Knuth's theoretical analysis. Ah, the power of a random number generator!

---

gs.m

```
01   G = 1/6;
02   n = input('How many elevators?');
03   totalgoingdown = 0;
04   elevator = zeros(n,4);
05   for loop = 1:1000000
06       for j = 1:n
07           decision(j) = rand;
08           if decision(j) < 0.5
09               elevator(j,1) = 0;
10           else
11               elevator(j,1) = 1;
12           end
```

(continued)

(continued)

```
13              elevator(j,2) = rand;
14              if elevator(j,2) < G
15                  if elevator(j,1) == 0
16                      elevator(j,3) = G + elevator(j,2);
17                  else
18                      elevator(j,3) = G − elevator(j,2);
19                  end
20                  elevator(j,4) = 1;
21              else
22                  if elevator(j,1) == 0
23                      elevator(j,3) = elevator(j,2) − G;
24                  else
25                      elevator(j,3) = (11/6) − elevator(j,2);
26                  end
27                  elevator(j,4) = 0;
28              end
29          end
30          min = elevator(1,3);
31          index = 1;
32          for k = 2:n
33              if elevator(k,3) < min
34                  min = elevator(k,3);
35                  index = k;
36              end
37          end
38          if elevator(index,4) == 0
39              totalgoingdown = totalgoingdown + 1;
40          end
41      end
42      totalgoingdown/1000000
```

6. Steve's Elevator Problem

For the $k = 2$ analysis (the riders at floor G are Steve and two others), we can write

- there is one stop for Steve if

    a. both of the other riders get off on Steve's floor (probability $= \frac{1}{n^2}$), or

    b. one of the other riders (either one) gets off on Steve's floor and the other gets off on a floor above Steve's (probability $= 2 \times \frac{1}{n} \times \frac{2}{n} = \frac{4}{n^2}$), or

    c. both of the other riders get off on a floor above Steve's (probability $= \frac{2}{n} \times \frac{2}{n} = \frac{4}{n^2}$);

- there are two stops for Steve if

    a. one of the other riders (either one) gets off on Steve's floor and the other gets of on any floor below Steve's (probability $= 2 \times \frac{1}{n} \times \frac{n-3}{n} = \frac{2(n-3)}{n^2}$), or

    b. both of the other riders get off on the same floor below Steve's (probability $= \frac{n-3}{n} \times \frac{1}{n} = \frac{n-3}{n^2}$), or

    c. one of the other riders (either one) gets off on a floor below Steve's and the other gets off on a floor above Steve's (probability $= 2 \times \frac{n-3}{n} \times \frac{2}{n} = \frac{4(n-3)}{n^2}$);

- there are three stops for Steve if

  a. each of the two other riders gets off on a different floor below Steve's $\left(\text{probability} = \frac{n-3}{n} \times \frac{n-4}{n} = \frac{(n-3)(n-4)}{n^2}\right)$.

Therefore, the average number of stops for Steve, for $k = 2$, is

$$1 \times \left[\frac{1}{n^2} + \frac{4}{n^2} + \frac{4}{n^2}\right] + 2 \times \left[\frac{2(n-3)}{n^2} + \frac{n-3}{n^2} + \frac{4(n-3)}{n^2}\right]$$

$$+ 3 \times \left[\frac{(n-3)(n-4)}{n^2}\right].$$

With a little algebra—which I'll let you verify—this reduces to $3 - \frac{7}{n} + \frac{3}{n^2}$. For Steve's building, with $n = 11$ floors, this becomes $3 - \frac{7}{11} + \frac{3}{121} = 2.3884$, which is in good agreement with the simulation results given in the problem statement. You can now appreciate why I didn't do similar analyses for $k \geq 3$; keeping track of all the possible ways that Steve's elevator companions could exit was just too complicated! But that's a moot issue for the Monte Carlo code steve.m.

S is the variable that, at the completion of one million elevator rides, will equal the total number of stops experienced by Steve. Line 01 initializes S to zero, and line 02 sets the value of k, the number of riders at floor G in addition to Steve. Lines 03 and 15 define the for/end loop that cycle steve.m through the million rides, while lines 04 through 14 simulate an individual ride. The heart of the code is the 11-element vector x; x(j) = 0 if the elevator did not stop at floor j, while x(j) = 1 if the elevator did stop at floor j. So, line 04 initially sets x(j) = 0, $1 \leq j \leq 11$, and then line 05 immediately sets x(9) = 1 because floor 9 (two floors below the top floor) is Steve's floor and we know he always gets off there. The question now is, where do the other k riders get off? That question is answered by lines 06 through 09, which for each of those riders randomly picks a floor from 1 to 11 (the value of the variable rider) and sets x(rider) = 1. This is done with the aid of the MATLAB command floor, which is a truncation command, e.g., floor(6.379) = 6 and floor(0.0931) = 0. Since 0<rand<1, then 0<11*rand<11, and so floor(11*rand) is an integer in the set (0, 1, 2, ..., 10). Therefore, rider is an integer in the set (1, 2, 3, ..., 11). Once x(j) is set equal to 1 for a particular value of j = j*, then of course, other riders who may also exit at that floor have no additional impact—x(j*) just gets set equal to 1 again.

The variable stops counts the number of floors at which the elevator stopped on its way up to Steve's floor; any stops above Steve's floor are of no interest. So, line 10 sets stops to 1 because we know x(9) = 1, and then lines 11 through 13 count the stops that happened from floor 1 to floor 8. Line 14 updates S, and then another ride is simulated. Line 16 gives us the average number of stops experienced by Steve on his morning ride to work after the one million rides have been completed.

---

steve.m

```
01    S=0;
02    k=input('How many riders in addition to Steve?');
03    for loop=1:1000000
04        x=zeros(1,11);
05        x(9)=1;
06        for j=1:k
07            rider=floor(11*rand)+1;
08            x(rider)=1;
09        end
10        stops=1;
11        for j=1:8
12            stops=stops+x(j);
13        end
14        S=S+stops;
15    end
16    S/1000000
```

When **steve.m** was run it produced the table given in the problem statement, as well as the additional values asked for:

Number of riders	Average number of stops
6 ($k = 5$)	4.03254
7 ($k = 6$)	4.48538
8 ($k = 7$)	4.894956
9 ($k = 8$)	5.267101
10 ($k = 9$)	5.608122

There is an amusing epilogue to the story of this problem. On April 4, 2005, I gave an invited talk to the Junior Colloquium in the mathematics department at the University of Tennessee, Knoxville. As an example of the intersection of pure mathematics and computers, I decided to discuss Steve's Elevator Stopping Problem and included a description of it in the advance announcement (which was placed on the Web) of my talk. A few days before the talk I received an e-mail from Michel Durinx, a member of the Theoretical Evolutionary Biology program at Leiden University in the Netherlands. Surprised as I was that someone in the Netherlands knew of my talk in Knoxville, I was even more surprised to see that Michel had included a complete theoretical solution to the problem. It's a pretty complicated (no surprise there!) combinatorial analysis, and the answer is that the average number of stops for Steve is $9 - 8(\frac{10}{11})^k$. This is such a simple expression that, as Michel himself wrote, it "suggests that with a fresher head, this result can be found [in a more elegant way]."

For the cases to which I already knew the theoretical answers ($k = 0$, 1, and 2), Michel's formula is in perfect agreement. For $k \geq 3$, his formula is 'confirmed' by the simulation results, i.e.,

k	Michel's formula
0	1
1	1.727272
2	2.388429
3	2.989481
4	3.535892
5	4.032629
6	4.484208
7	4.894735
8	5.267940
9	5.607219

You can compare this table with the entries in the table from the problem statement ($0 \leq k \leq 4$) and the table I just gave ($5 \leq k \leq 9$). And just why was Michel in the Netherlands reading announcements on the Web from a math department in Tennessee? Well, actually he wasn't.

What happened is that a colleague of his at Leiden, Frans Jacobs, who had had a postdoctoral appointment at the University of Tennessee, was still on the department's math server. Frans got the announcement of my talk, shared it with his friend Michel, and the result was that I got a solution. Ah, the power of the Web!

7. The Pipe Smoker's Discovery

The code smoker.m simulates the problem one million times, as controlled by the for/end loop defined by lines 03 and 16. Before beginning those simulations, however, line 01 initializes the variable S to zero, where the value of S will, at the completion of each simulation, be the current total number of matches used (and so, when the program completes the last of the one million simulations, S/1000000, as computed in line 17, will be the answer to the original question, i.e., the average number of matches removed until one booklet is empty). Also, line 02 initially sets all eighty elements of the row vector matches to zero, where matches(x) will, at the completion of each simulation, be the current number of times that the number of removed matches was equal to x. You should see that $40 \leq x \leq 79$, always. The "79" occurs if the smoker has happened to pick booklets such that he eventually arrives at the situation with each booklet having just one match left. At this point, 78 matches have of course been removed. The next selection will then remove the 79th match, leave one of the booklets empty, and so immediately terminate the process. The case of "40" occurs, obviously, if the smoker selects the same booklet 40 times in a row. Lines 03 through 16 are the heart of the program, carrying out the logic of each individual one of the one million simulations. In line 04 the program sets the initial number of matches in the two booklets

---

smoker.m

```
01   S = 0;
02   matches = zeros(1,80);
03   for k = 1:1000000
04       booklet1 = 40;booklet2 = 40;
05       while booklet1 > 0&booklet2 > 0
06           r = rand;
07           if r < 0.5
08               booklet1 = booklet1 − 1;
09           else
10               booklet2 = booklet2 − 1;
11           end
12       end
13       s = (40 − booklet1) + (40 − booklet2);
14       S = S + s;
15       matches(s) = matches(s) + 1;
16   end
17   S/1000000
18   bar(matches)
```

(booklet1 and booklet2) to 40. Line 05 controls how long each of the individual simulations takes; it will, in general, be different from one simulation to the next, depending on just how the smoker happens to randomly select booklets. Line 05 says to keep going as long as both booklets have at least one match left; as soon as one of them (either one, it doesn't matter which) is empty, then the while statement will terminate the looping. As long as both match booklets are nonempty, however, the looping (i.e., match removal) continues. Line 06 generates a random number r from a distribution that is uniform from 0 to 1, and lines 07 through 11 say that, with equal probability (of 0.5), the smoker takes his match from one or the other of the two booklets. When the while statement in line 05 finally detects that one of the two booklets is empty, the simulation loop is exited and the variable s is computed; (40–booklet1) and (40–booklet2) are the number of matches that have been removed from booklet1 and booklet2,

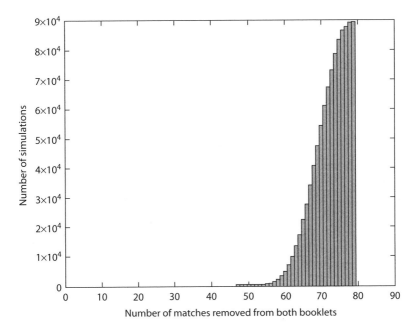

Figure S7.1. The pipe smoker's problem.

respectively, and so line 13 computes the total number of matches removed. Obviously, one of these two expressions in parentheses is zero; i.e., the booklet that is empty has had all of its matches removed by definition, but we don't need to know which one it is. With s computed, line 14 uses it to update S (the total number of matches removed to date), and also to update the vector matches(s) that's keeping track of the number of simulations that have s matches removed. With those two bookkeeping operations done, smoker.m then starts another simulation until, after one million of them, the for/end loop defined by lines 03 and 16 is exited and line 17 produces the average number of matches removed per simulation. In line 18 the MATLAB function bar generates a bar graph plot of matches—see Figure S7.1. (I have not shown, in the coding for smoker.m, the program statements that generate the figure labeling.)

The average number of matches removed from both booklets until one of them was empty was reported by smoker.m to be 72.869, with the range of values on the number of matches removed being 47 to

79, with the 47 value occurring just once. We shouldn't be surprised that we didn't see any occurrences of the number of removed matches less than than 47, even after a million simulations. Such small numbers of removed matches represent very low-probability events; e.g., the probability of removing the smallest possible number of matches (40) is just $2(1/2)^{40} \approx 1.82 \times 10^{-12}$ (the first factor of 2 is because either of the two booklets could be the one that ends up as the empty booklet). Higher numbers of removed matches, however, are quite likely, as illustrated in Figure S7.1. For example, smoker.m reported the number 79 a total of 89,303 times, giving a probability of 0.0893.

The authors of the 1972 book on my bookshelf didn't tell their readers that this smoker's problem is a variation on a famous probability problem called the Banach Matchbox Problem. Named after the Polish mathematician Stefan Banach (1892–1945), the problem was the creation of Banach's close friend and fellow Pole Hugo Steinhaus (1887–1972), who, at a conference in Banach's honor, posed it as follows:

> A mathematician who loved cigarettes (Banach smoked up to five packs a day and his early death at age 53 was, not surprisingly, of lung cancer) has two matchboxes, one in his left pocket and one in his right pocket, with each box initially containing N matches. The smoker selects a box at random each time he lights a new cigarette, and the problem is to calculate the probability that, when the smoker first *discovers* that the box he has just selected is empty, there are exactly r matches in the other box, where clearly $r = 0, 1, 2, \ldots, N$.

The reason for emphasizing the word *discovers* is that we are to imagine that when the smoker takes the last match out of a box, he puts the box back into the pocket it came from, and he only becomes aware that the box is empty when he next selects that box. This isn't as stupid as it may sound on first encounter, if we imagine that the smoker doesn't actually look in a box as he removes a match but rather is so engrossed in the deep mathematics he is scribbling on paper that he simply gropes around inside the box with his fingers until he feels a match (or finally realizes—i.e., *discovers*—that the box is in fact empty).

Banach's problem, then, is not quite the same as our original smoker's problem. In the original problem the booklet selection process immediately stops at the moment one of the booklets is exhausted. In Banach's problem the selection process continues until the empty box is once again selected. Therefore, one or more additional matches may be removed from the nonempty box before the smoker again selects the empty box. Indeed, in Banach's problem it is perfectly possible for both boxes to end up empty (the above $r = 0$ case)! What does this new twist do to our simulation code smoker.m? Actually, not too much, with the new smokerb.m being the original smoker.m with some additional code inserted between lines 12 and 13—that is, immediately after the while loop. Here's how it works.

All goes as before, but now when the while loop is exited smokerb.m executes the new line 13, which sets the variable empty to zero, which in turn will control the operation of a new, second while loop defined by lines 14 and 29. The meaning of empty = 0 is that the booklet that has just been emptied has not yet been reselected. Line 15 randomly selects one of the booklets; lines 16 through 21 (executed if the selected booklet is booklet1) takes a match from that booklet if it still contains at least one match; otherwise, no match is taken and empty is set equal to 1. Lines 22 through 27 do the same for booklet2 if it is the selected booklet. The first time the code encounters an empty booklet the resulting empty = 1 condition terminates the while loop and, then all goes as before, as it did in smoker.m.

When run, smokerb.m produced Figure S7.2, with the estimate 73.79 for the average number of removed matches, about 1 greater than for the original problem. As Figure S7.2 illustrates, the possibility now exists for both booklets to be exhausted, as I mentioned before.

We can analytically check these simulation results as follows, where here I'll use the dental floss version of the problem. (Is the dental floss problem a version of the original problem, which terminates as soon as one of the boxes is empty, or is it a form of Banach's problem, which terminates when the empty box is discovered on its next selection to be empty? I think the argument for the original problem is stronger, as one knows immediately when the floss runs out!)

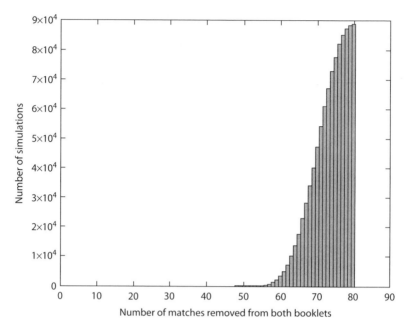

Figure S7.2. Banach's problem.

As the author of that problem wrote,[1]

The probability of finding k ft of floss in a given box when the
other box *becomes* [my emphasis, and this is another clue for
the dental floss problem not being the Banach problem] empty
is the same as the probability of finding $N - k$ tails in a sequence
of random [fair] coin tosses with exactly N heads, when the last
toss is a head and $N = 150$. The probability $P(k)$ of finding k ft of
floss in either box is twice this, since either box can run out with
equal probability. The probability that the last toss is a head is
$1/2$, and so $P(k)$ is just the probability of $N - 1$ heads and $N - k$
tails in a sequence of random tosses [of length $2N - k - 1$]. This
is given by the binomial distribution; explicitly

$$P(k) = \left(\tfrac{1}{2}\right)^{2N-k-1} \frac{(2N-k-1)!}{(N-1)!(N-k)!}.$$

The expectation value of the length of remaining floss is $\langle k \rangle$
$= \sum_{k=1}^{N} k P(k)$; and, for $N = 150$, $\langle k \rangle = 13.8\,\text{ft}$.

smokerb.m

```
01   S = 0;
02   matches = zeros(1,80);
03   for k = 1:1000000
04       booklet1 = 40;booklet2 = 40;
05       while booklet1 > 0&booklet2 > 0
06           r = rand;
07           if r < 0.5
08               booklet1 = booklet1 − 1;
09           else
10               booklet2 = booklet2 − 1;
11           end
12       end
13       empty = 0;
14       while empty == 0
15           r = rand;
16           if r < 0.5
17               if booklet1 > 0
18                   booklet1 = booklet1 − 1;
19               else
20                   empty = 1;
21               end
22           else
23               if booklet2 > 0
24                   booklet2 = booklet2 − 1;
25               else
26                   empty = 1;
27               end
28           end
29       end
30       s = (40 − booklet1) + (40 − booklet2);
31       S = S + s;
32       matches(s) = matches(s) + 1;
33   end
34   S/1000000
35   bar(matches)
```

Just to clarify a bit what all of the above means: if we flip a fair coin $2N - k$ times, and if we observe that the coin shows heads N times (heads means we use the floss box that will eventually be the empty one) and shows tails $N - k$ times (tails means we use the floss box that will, when the other box becomes empty, still have k feet left in it), then the probability of this happening is the probability that there are k feet of floss left in the nonempty box. And, as explained above,

$$P(k) = (2)\left(\tfrac{1}{2}\right)\left[\binom{2N-k-1}{N-k}\left(\tfrac{1}{2}\right)^{2N-k-1}\right],$$

where the first factor of 2 is there because either box could be the one that ends up empty, the second factor of $(\tfrac{1}{2})$ is the probability that the last toss is a heads, and the final factor in braces is the probability of $N - k$ tails in the first $2N - k - 1$ tosses. So,

$$P(k) = \left(\tfrac{1}{2}\right)^{2N-k-1}\binom{2N-k-1}{N-k}$$

as claimed. And, again as claimed,

$$\langle k \rangle = \sum_{k=1}^{N} k\, P(k) = \sum_{k=1}^{N} k \left(\tfrac{1}{2}\right)^{2N-k-1}\binom{2N-k-1}{N-k}.$$

This is easy to code in MATLAB using the nchoosek command to evaluate the binomial coefficients—nchoosek(x,y)$= \binom{x}{y}$ for x and y non-negative integers. The code floss.m does the job, with the final value of sum being 13.8 (as claimed) for $N = 150$.

In fact, we can use floss.m to check one last claim. As the author of the problem wrote, "In general, $\langle k \rangle \simeq \sqrt{N}$," which means $\langle k \rangle$ increases as $\sqrt{N}$, i.e., as $N^{1/2}$. In other words, $\langle k \rangle \simeq C N^{1/2}$, where C is some constant[2], and so $\log \langle k \rangle = \log C N^{1/2} = \log C + 1/2 \log N$. This means that if we do a log-log plot of $\langle k \rangle$ versus N, we should see a straight line with slope 1/2. Figure S7.3 shows a log-log plot of $\langle k \rangle$ versus N as N varies from 1 to 500, and it is indeed straight (at least to the eye). To confirm that the slope is 1/2, I've also plotted (as a dashed line) the reference line with $10\sqrt{N}$ on the vertical axis (the 10 is there simply

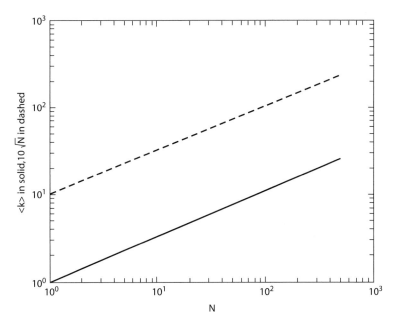

Figure S7.3. The Dental Floss Problem.

---

floss.m

```
01    sum = 0;
02    N = 150;
03    for k = 1:N
04        sum = sum + (k*nchoosek(2*N−k−1,N−k)/(2^(2*N−k−1)));
05    end
06    sum
```

---

to offset this reference line from the solid line), and you can see that the two lines are indeed parallel (at least to the eye) and so have the same slope.

For our original match booklet problem we simply run floss.m for $N = 40$ and find that $\langle k \rangle = 7.114$ matches left in the nonempty booklet. This is in good agreement with the estimate from smoker.m of 72.872 matches removed (from the initial 80) when one of the booklets becomes empty ($80 - 72.872 = 7.128$).

One last comment: Why is the dental floss problem called a paradox? The answer lies in understanding the difference between the *average value* of k and the *most probable value* of k. They are very different. Here's why. Notice that

$$\frac{P(k+1)}{P(k)} = \frac{\left(\frac{1}{2}\right)^{2N-(k+1)-1} \dfrac{(2N-(k+1)-1)!}{(N-1)!(N-(k+1))!}}{\left(\frac{1}{2}\right)^{2N-k-1} \dfrac{(2N-k-1)!}{(N-1)!(N-k)!}}$$

$$= \left(\frac{1}{2}\right)^{-1} \times \frac{(2N-k-2)!}{(2N-k-1)!} \times \frac{(N-1)!(N-k)!}{(N-1)!(N-k-1)!}$$

$$= 2 \times \frac{1}{2N-k-1} \times (N-k)$$

$$= \frac{2N-2k}{2N-k-1},$$

which is clearly less than 1 for all $k > 1$ (and this ratio steadily decreases as k increases). But, for $k = 1$, we have

$$\frac{P(2)}{P(1)} = \frac{2N-2}{2N-2} = 1,$$

and so $P(1) = P(2) > P(3) > P(4) > \cdots$. Notice that this is true for any value of N. The most probable values of k, for any value of N, are the equally probable $k = 1$ feet and $k = 2$ feet, and all other values of k (in particular, $\langle k \rangle$) are steadily less probable.

References and Notes

1. Peter Palffy-Muhoray, "The Dental Floss Paradox" (*American Journal of Physics*, October 1994, pp. 948 and 953).
2. The exact expression for the Banach problem is $\langle k \rangle = \frac{2N+1}{2^{2N}}\binom{2N}{N} - 1$, which, using Stirling's formula to approximate the factorials, reduces to $2\sqrt{\frac{N}{\pi}} - 1$. You can find this worked out (it is not easy to do!) in the second edition of William Feller, An Introduction to Probability Theory and Its Applications (New York; John Wiley & Sons, 1957, vol. 1, pp. 212–213). For

some reason Feller omitted this calculation in the third edition. For $N = 40$, the approximation gives $\langle k \rangle = 6.14$, which agrees well with the estimate from smokerb.m for the number of matches removed (from the initial 80), i.e., $80 - 73.79 = 6.21$. The exact expression gives $\langle k \rangle = 6.20$, which is in even better agreement with smokerb.m.

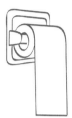

8. A Toilet Paper Dilemma

The code rolls.m uses Knuth's recurrence formulas to directly calculate and plot $M_{200}(p)$. This program can be most easily understood by referring to Figure S8.1, which again shows the underlying lattice point structure of the problem. The one complication MATLAB itself, as a programming language, introduces is that it does not allow zero indexing of arrays. If your favorite language does, then rolls.m will still work, but you could, if you wanted, actually rewrite the program with fewer commands. As you can see, however, rolls.m is pretty short anyway.

Line 01 starts things off by defining M to be a 200×200 array. The variable prob, initially set equal to zero in line 02, will be the index into the row vector curve, where curve(prob) will be the value of $M_{200}(p)$, with p=prob/1000, as in line 05. Here's how that works. rolls.m computes $M_{200}(p)$ for $0.001 \le p \le 1$, in steps of 0.001, and so lines 03 and 20 define a for/end loop that will cycle a thousand times, each time using p=prob/1000 as defined in line 05. Notice that $p = 0.001$ goes with prob=1, that $p = 0.002$ goes with prob=2, and so on, because line 04 immediately increments prob by 1, to 1, from its initial value of 0 (remember, no zero indexing in MATLAB). Since, as explained in the problem statement, $M(1, 1) = 1$ for any n and any p, line 06 does precisely that. Lines 07 through 09 are a loop that is present because of the zero indexing issue; looking at each lattice point in Figure S8.1, the value of M at any given point is, by Knuth's recurrence (d), the weighted sum of the M-value at the lattice point to the immediate left

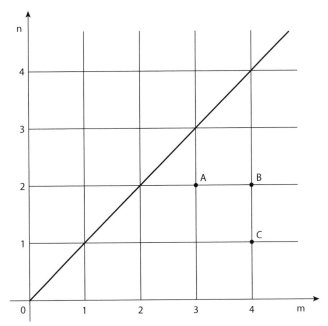

The value of M at point A is the weighted sum of the M values at point B and point C for all lattice points *below* the diagonal line.

Figure S8.1. The lattice point structure of the toilet paper problem.

and at the lattice point directly below the given point. But of course, for row 1 of M, all the points directly below that row would be in row 0—which is an impossible row in MATLAB. So, line 08 directly calculates $M(m, 1)$ not as $(1 - p)^* M(m, 0) + p^* M(m - 1, 1)$ but rather as shown in rolls.m, with $M(m, 0)$ replaced by m, as specified in Knuth's recurrence (b). Line 10 then completes the initial part of rolls.m by using Knuth's recurrence (c) to set the diagonal lattice point M-value at $(2, 2)$, i.e., $M(2, 2) = M(2, 1)$.

rolls.m now computes the value of M at all the remaining lattice points in a vertical pattern. That is, for every column in M, starting with the third column, the code computes M at each lattice point up to the lattice point just below the lattice point on the diagonal. The lattice point on the diagonal is assigned the same value given to the lattice point just below it, as required by Knuth's recurrence (c). This is all accomplished by lines 11 through 18. The end result is that the

---

rolls.m

```
01    M = zeros(200,200);
02    prob = 0;
03    for loop = 1:1000
04          prob = prob + 1;
05          p = prob/1000;
06          M(1,1) = 1;
07          for m = 2:200
08                M(m,1) = (1 − p)*m + p*M(m − 1,1);
09          end
10          M(2,2) = M(2,1);
11          diagonal = 3;
12          for m = 3:200
13                for n = 2:diagonal-1
14                      M(m,n) = p*M(m − 1,n) + (1 − p)*M(m,n − 1);
15                end
16                M(m,n) = M(m,m − 1);
17                diagonal = diagonal + 1;
18          end
19          curve(prob) = M(200, 200);
20    end
```

---

value of M at (200, 200), for the current value of p (equal to prob/1000) is arrived at and stored as the probth element in the curve vector.

When rolls.m was run (I have not included in the above listing the specialized MATLAB plotting and axis-labeling commands) it created Figure S8.2, and as you can see, the plot of $M_{200}(p)$ has a curious feature in it around $p = 0.5$. As Knuth writes, "The function $M_n(p)$ is a polynomial in p of degree $2n − 3$ [and so in Figure S8.2 we have a curve of degree 397], for $n \geq 2$, and it decreases monotonically from n down to 1 as p increases from 0 to 1. The remarkable thing about this decrease is that it changes in character rather suddenly when p passes 1/2."

I mentioned in the original statement of this problem that Knuth's paper has achieved a certain level of fame among computer scientists.

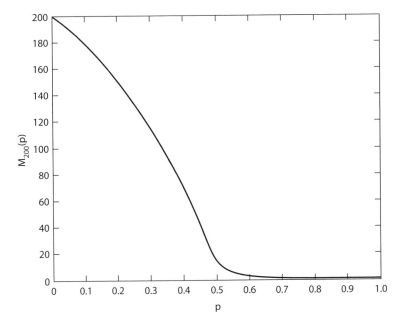

Figure S8.2. How the paper runs out.

This is because, in addition to discussing an interesting mathematical problem, it treats a potentially taboo topic in a totally professional manner—but I am sure that as he wrote his paper Knuth had at least the hint of a smile on his face. Others, however, have since allowed themselves to venture just a little further. For example, in one Web discussion of the problem that I once came across, Knuth's publication was called a "tongue-in-cheek paper." That same writer then suggested a generalization of the problem whose solution would "certainly fill an important gap." Who says computer scientists don't have a sense of humor? Warped it may be, yes, but nonetheless it is there. If he ever came across that Web site, I am sure Knuth would have let himself go (oh, my, now *I'm* making potty jokes, too!) and have laughed out loud. I know I did! Knuth doesn't mention it, but I do find it amusing to note the curious resemblance of the curve in Figure S8.2 with the actual appearance of a long length of loose toilet paper hanging from a roll onto a bathroom floor.

9. The Forgetful Burglar Problem

The code of fb.m simulates ten million paths of the burglar, with each path continuing until a previously visited house is revisited. Line 01 defines the row vector duration as having fifty elements, where duration(k) is the number of paths with length k steps (each step is a move of length 1 or 2). Lines 02 and 19 define the for/end loop that cycle fb.m through its ten million simulations. Each simulation will continue as long as the variable go is equal to 1, to which it is initially set in line 03. Line 04 initializes the row vector whereyouvebeen to all zeros, where the meaning of whereyouvebeen(k)=0 is that the burglar has (not yet) visited location k. I have somewhat arbitrarily made whereyouvebeen of length 201, and have initially placed the burglar in the middle of the vector; i.e., lines 05 and 06 force whereyouvebeen(101)=1. (This gives plenty of room for the burglar to wander around in, in either direction.) That is, the burglar's criminal paths of robbery have to, obviously, always start somewhere, and that place (and all subsequently visited places) are marked with a 1 in whereyouvebeen. Then line 07 initializes the variable steps to 0, the current length of the burglar's path of crime.

The while loop, defined by lines 08 and 18, now performs an individual simulation which will continue until, as explained above, the condition go = 0 occurs (you'll see how that happens soon). Line 09 probably needs some explanation. Since the burglar can move in either direction (with equal probability) by either one or two houses (again, with equal probability) from his current victim's house, we need to generate the current move as one of the equally probable numbers $-2, -1, 1,$ and 2. Since rand generates a number uniformly distributed between 0 and 1 (but never equal to either 0 or 1), then floor(2*rand) is equal to either 0 or 1 with equal probability. (Remember, as explained in the solution to Problem 6, Steve's Elevator Problem, floor is MAT-LAB's "rounding down" or truncation command.) Thus, floor(2*rand)+1 is equal to either 1 or 2 with equal probability. The command sign implements what mathematicians call the *signum* function, i.e., sign(x) = +1 if $x > 0$ and sign(x) = −1 if $x < 0$. Since rand−0.5 is positive and negative with equal probability, then sign(rand−0.5) is $+1$ and -1 with equal probability. Thus, line 09 generates the required $-2, -1, 1,$ and 2 with equal probability and assigns that value to the variable move.

```
fb.m
01    duration = zeros(1,50);
02    for loop = 1:10000000
03        go = 1;
04        whereyouvebeen = zeros(1,201);
05        hereyouare = 101;
06        whereyouvebeen(hereyouare) = 1;
07        steps = 0;
08        while go == 1
09            move = sign(rand-0.5)*(floor(2*rand) + 1);
10            hereyouare = hereyouare + move;
11            steps = steps + 1;
12            if whereyouvebeen(hereyouare) == 1
13                go = 0;
14                duration(steps) = duration(steps) + 1;
15            else
16                whereyouvebeen(hereyouare) = 1;
17            end
18        end
19    end
20    duration/10000000
```

Lines 10 and 11 then update hereyouare, the new location of the burglar, and the variable steps.

Line 12 is the start of an if/end loop that determines if the new location is a previously visited one; if so, line 13 sets go to zero, which will terminate the while loop, and line 14 records the length of the now terminated path of theft in the vector duration. If the new location has not been previously visited, then line 16 marks that location as now having been visited. The code continues in this fashion through all ten million simulated paths, and then, finally, line 20 produces the fb.m's probability estimates for all path lengths from 1 to 50.

The following table compares the estimates produced by fb.m with the theoretical probabilities of the forgetful burglar first returning to a previously burgled home in exactly k steps, $1 \leq k \leq 7$ (calculated from a rather complicated formula[1]).

k	Theoretical probability	fb.m's estimate
1	0	0
2	$\frac{4}{16} = 0.25$	0.2500623
3	$\frac{18}{64} = 0.28125$	0.2811653
4	$\frac{50}{256} = 0.1953125$	0.1954533
5	$\frac{120}{1,024} = 0.1171875$	0.1171681
6	$\frac{280}{4,096} = 0.0683594$	0.0683619
7	$\frac{638}{16,384} = 0.0389404$	0.0388911

This problem is a generalization of one type of a so-called *random walk with absorption*. In the traditional form, the one usually given in probability textbooks as a *gambler's ruin* problem, the steps to the left and to the right are the results of losses and wins, with the walk terminated if the gambler reaches either the far left location, where he has zero money and so is ruined, or the far right location, where he has won all the money and his opponent is ruined. In both cases the gambler is said to be *absorbed* at the location. In the forgetful burglar problem, each newly visited location *transforms* into an absorbing location because, if any location is visited a second time, the walk is immediately terminated. As far as I know, the general solution to this form of random walk is still unknown.

References and Notes

1. Caxton Foster and Anatol Rapoport, "The Case of the Forgetful Burglar" (*American Mathematical Monthly*, February 1958, pp. 71–76).

10. *The Umbrella Quandary*

The code umbrella.m performs 10,000 simulations of the man's walks for each of the ninety nine values of rain probability, from 0.01 to 0.99, in steps of 0.01. That is, it performs a total of 990,000 simulated walks. We don't have to simulate for $p = 0$ (it never rains) because, obviously, in that case the man never gets wet; i.e., the answer for $p = 0$ is that he'll walk an infinite number of times before his first soaking. That's the answer for $p = 1$, too (it always rains), but for a far different reason. For $p = 1$ the man always carries an umbrella with him and so, again, he never gets wet. These two cases of deterministic rain would get a simulation in trouble, in any case: the looping would continue forever while the code waited for the first soaking to occur! Where our Monte Carlo simulation comes into its own is, naturally, for the far more interesting cases of probabilistic rain.

Lines 01 and 02 bring in the values of the number of umbrellas that the man starts with at home (xi) and at the office (yi). Line 03 defines the 99-element row vector duration, where duration (k) will, on completion of program execution, be the average number of walks before the man's first soaking for the rain probability $k/100$; e.g., duration(17) will be the answer to our problem for $p = 0.17$. Lines 04 and 43 define the outermost for/end loop, a loop that will be executed ninety nine times—once for each value of p from 0.01 to 0.99 in steps of 0.01; line 05 sets the value of the rain probability p. Line 06 initializes the variable walksum to zero; walksum will be the total number of walks, over 10,000 simulations of the problem with a given p, before the man gets wet. (The variable walks, initialized at zero in line 10, is the number of such walks for a single simulation.) Lines 07 and 41 define the for/end loop that runs umbrella.m through the 10,000 simulations for the current value of p. Lines 08 and 09 set x and y at the start of each simulation to xi and yi, and line 10 initializes walks as stated before.

Line 11 needs a little elaboration. To keep track of where the man is at all times, I've coded "home" as -1 and "office" as 1. Thus, to move the man back and forth, all we need do to get his new location is to multiply the old location by -1 (because $-1 \times -1 = 1$ and $1 \times -1 = -1$). Since the original problem statement said the man always starts at home, then the variable location is initially set equal to -1. Lines 13 and 40 define a while loop that controls the individual simulations. This loop is executed an a priori undefined number of times (until the man gets wet), and it is controlled by the variable wet; wet=0 means the man is dry, and wet=1 means (surprise!) that he is wet, the condition that terminates the while loop.

---

umbrella.m

```
01      xi = input('Initial umbrellas at home?');
02      yi = input('Initial umbrellas at office?');
03      duration = zeros(1,99);
04      for P = 1:99
05          p = P/100;
06          walksum = 0;
07          for loop = 1:10000
08              x = xi;
09              y = yi;
10              walks = 0;
11              location = - 1;
12              wet = 0;
13              while wet == 0
14                  if rand > p
15                      walks = walks + 1;
```

(continued)

(continued)

```
16                              location = − 1*location;
17                         else
18                              if location == − 1
19                                  if x == 0
20                                       walksum = walksum + walks;
21                                       wet = 1;
22                                  else
23                                       x = x − 1;
24                                       walks = walks + 1;
25                                       y = y + 1;
26                                       location = − 1*location;
27                                  end
28                              else
29                                  if y == 0
30                                       walksum = walksum + walks;
31                                       wet = 1;
32                                  else
33                                       y = y − 1;
34                                       walks = walks + 1;
35                                       x = x + 1;
36                                       location = − 1*location;
37                                  end
38                              end
39                         end
40                     end
41               end
42               duration(P) = walksum/10000;
43         end
```

The while loop starts by determining (line 14) if it is raining when the man starts his current walk. If it is not raining, then lines 15 and 16 simply increment the number of dry walks by one and move the man to the other location. If it is raining then either the block of lines from 18 through 26 is executed, or the block of lines from 29 through 36 is; which block is executed depends on where the man is. If he is at

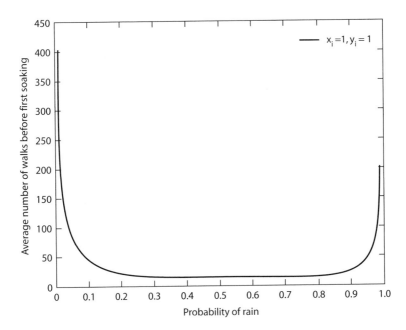

Figure S10.1. The umbrella problem (plotted for $x_i = y_i = 1$).

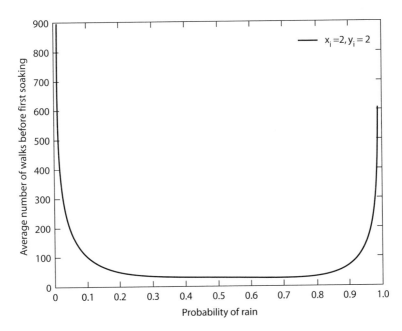

Figure S10.2. The umbrella problem (plotted for $x_i = y_i = 2$).

home (if location=−1) then the code checks if there is a home umbrella available (line 19). If there is not, then line 20 updates walksum and line 21 sets wet equal to 1, which will terminate that particular simulation, i.e., will terminate the while loop. If there is a home umbrella available, then x is decremented by one (line 23), walks is incremented by one (line 24), y is incremented by one because the man has transported an umbrella to his office during his walk (line 25), and the man's location is updated (line 26). If, on the other hand, the man is at his office (location=1) then the second block of code does all the above but with the obvious changes, e.g., if an umbrella is transported, then it is y that is decremented and x that is incremented. When the 10,000 simulations are done for the current value of p, line 42 updates duration, and then another 10,000 simulations are performed for the next value of p.

Figures S10.1 shows a plot of duration vs. p for xi=yi=1, and Figures S10.2 shows the plot for xi=yi=2. (I have not included the MATLAB plotting and labeling commands in umbrella.m.) Both plots illustrate the expected behavior in the limiting cases as $p \to 0$ and $p \to 1$. For how to treat this problem analytically, see any good book on the theory of Markov chains.

11. The Case of the Missing Senators

The code missing.m simulates the problem of the absentee senators. Lines 01 and 02 are by now obvious to you. Line 03 initializes the variable defeats to zero; the value of this variable, at the end of the program execution, will be how many times the bill was defeated during one million simulated votes. Lines 04 and 26 define the for/end loop that cycles the code through its one million simulations. Line 05 defines the 100-element row vector votes; votes(k) is the vote during an individual simulation of the kth senator (if that senator actually does vote), where -1 is a vote against the bill and 1 is a vote for the bill. This explains the two for/end loops, in lines 06 through 08 and again in lines 09 through 11. Lines 12 through 21 implement the logic of randomly removing M of the senators from the vote. The key—and easy to overlook, I think—observation is that we must be careful to eliminate M different senators. That is, once a senator has been eliminated, the code must not be allowed to later randomly select that same senator for removal!

The net result, when the code reaches line 22, is that the senators who show up for the vote are represented in the votes vector by either a -1 (against) or a 1 (for), and the senators who miss the vote are represented in votes by a 0. So, we can get the result of the vote by simply summing the elements of votes, a calculation performed in line 22 with MATLAB's nifty sum command; the result is assigned to the variable vote. Clearly, vote > 0 means the bill passed and vote < 0 means the bill failed (and vote $= 0$ means a tie vote). This explains lines 23 through 25. Line 27, executed after one million voting simulations, generates missing.m's answer. When run for the check case I gave you in the original problem statement ($A = 49$ and $M = 3$), missing.m produced an estimate for the probability that the bill is defeated of 0.128492. This is in good agreement with the theoretical answer of $0.128787\cdots$. Now, before giving you the Monte Carlo results for the remaining questions I asked you to consider, let me show you how to calculate the theoretical answers.

If we define x as the number of senators missing the vote that, if present, would have voted for the bill, then $M - x$ is the number of senators missing the vote who would have voted against the bill. Now, the bill will be defeated if $(100 - A) - x < A - (M - x)$, or, after some rearranging, the bill will be defeated if $x > 50 - A + \frac{M}{2}$. Obviously, senators come only in integers, and since M could be an odd integer, then we more precisely write the condition for the bill being defeated as $x \geq [50 - A + \frac{M}{2}] + 1$, where the notation $[y]$ means the largest integer less than or equal to y. For example, $[2.5] = 2$ and $[3] = 3$.

---

missing.m

```
01    A = input('Number of senators against bill?');
02    M = input('Number of senators missing vote?');
03    defeats = 0;
04    for loop = 1:1000000
05          votes = zeros(1,100);
06          for k = 1:A
07                votes (k) = -1;
08          end
09          for k = A + 1:100
10                votes(k) = 1;
11          end
12          for k = 1:M
13                go = 1;
14                while go == 1
15                      j = floor(100*rand) + 1;
```

(continued)

(continued)

```
16                    if votes(j)~ = 0
17                        votes(j) = 0;
18                        go = 0;
19                    end
20                end
21            end
22            vote = sum(votes);
23            if vote < 0
24                defeats = defeats + 1;
25            end
26        end
27    defeats/1000000
```

The probability the bill is defeated is

$$\text{Prob}\left(x > 50 - A + \frac{M}{2}\right) = \sum_{k=\left[50-A+\frac{M}{2}\right]+1}^{M} \text{Prob}(x = k).$$

The number of ways to pick k senators to miss the vote, from those who would have voted for the bill, is $\binom{100-A}{k}$, and the number of ways to pick the other $M - k$ senators who miss the vote, from those who would have voted against the bill, is $\binom{A}{M-k}$, and so the total number of ways to pick the M missing senators, with exactly k of them being for the bill *but such that the bill is defeated*, is $\binom{100-A}{k}\binom{A}{M-k}$. And since the total number of ways to pick the M missing senators *without any constraints* is $\binom{100}{M}$, we have

$$\text{Prob}(x = k) = \frac{\binom{100-A}{k}\binom{A}{M-k}}{\binom{100}{M}},$$

i.e.,

$$\text{Probability the bill is defeated} = \sum_{k=\left[50-A+\frac{M}{2}\right]+1}^{M} \frac{\binom{100-A}{k}\binom{A}{M-k}}{\binom{100}{M}}.$$

For the "check your code" case of $A = 49$ and $M = 3$, this probability is

$$\sum_{k=3}^{3} \frac{\binom{100-49}{k}\binom{49}{3-k}}{\binom{100}{3}} = \frac{\binom{51}{3}\binom{49}{0}}{\binom{100}{3}} = \frac{\frac{51!}{3!48!}}{\frac{100!}{3!97!}}$$

$$= \frac{51 \times 50 \times 49}{100 \times 99 \times 98} = \frac{51}{396} = 0.12878787\cdots,$$

as given in the original problem statement.

I'll leave it for you to confirm that the above theoretical analysis gives the following probabilities to the two other cases you were to consider: $\frac{\binom{51}{4}}{\binom{100}{4}} = 0.06373$ (for $A = 49$ and $M = 4$), and $\frac{\binom{51}{4}\binom{49}{1} + \binom{51}{5}\binom{49}{0}}{\binom{100}{5}}$ $= 0.193845\cdots$ (for $A = 49$ and $M = 5$). When run, missing.m gave the following estimates: 0.063599 and 0.193939, respectively. Okay!

12. How Many Runners in a Marathon?

The code estimate.m simulates this problem. It is probably the most MATLABy of all the codes in this book, with its use of several specialized (but all quite easy to understand), highly useful vector commands. All of these commands could be coded in detail—as I did in Problem 3 with the bubble-sort algorithm, rather than just using MATLAB's sort command—but that would be tedious. If your favorite language doesn't have similar commands, I'll let the tedium be yours! And in estimate.m, since the code is so MATLABy anyway, I've made an exception to my earlier decisions not to include the plotting and figure labeling commands. All such commands in estimate.m are included here for your inspection. Okay, here's how the code works.

Line 01 defines the four element values of the vector size to be the four sample size percentages for which you were asked to generate performance histograms. For each such value, estimate.m will simulate 10,000 populations of randomly selected sizes, and the outermost for/end loop defined by lines 02 and 29 cycle the code through those four sample sizes. Line 03 sets the variable s equal to the current value of the sample size percentage. Lines 04 and 22 define the for/end loop that performs each of those 10,000 population simulations for a fixed sample size percentage. Line 05 randomly picks an integer value from the interval 100 to 1,000 for the population size N. (The MATLAB command round rounds its argument to the nearest integer, unlike the MATLAB command floor, which truncates its argument, i.e., rounds downward, and the command ceil, which rounds upward, "toward the ceiling.") Line 06 uses the nifty MATLAB command randperm to create the vector population, whose elements are a random permutation of the integers 1 through N (you'll remember we used randperm in the code guess.m in the introduction; it's also used in the code optimal.m in the solution to Problem 20). It is population from which estimate.m will be sampling without replacement. Line 07 then uses N and s to set the integer variable n to the actual sample size. To prepare for that sampling of population, the variables newN and newn are initialized to the values of N and n, respectively, in line 08. And finally, line 09 defines the vector observed, which will store the n values from population that are generated by the sampling process.

Lines 10 through 19 are a direct implementation of Bebbington's sampling without replacement algorithm. Once the n samples of population have been taken and stored in observed, the Bebbington algorithm is exited, and line 20 assigns the largest value found in observed to the variable maximum (with the aid of MATLAB's max command).

---

estimate.m

```
01      size(1) = 2;size(2) = 5;size(3) = 10;size(4) = 20;
02      for z = 1:4
03          s = size(z);
04          for loop = 1:10000
05              N = round(900*rand) + 100;
06              population = randperm(N);
07              n = round(s*N/100);
08              newN = N;newn = n;
09              observed = zeros(1,n);
10              j = 1;
11              for k = 1:N
12                  p = newn/newN;
13                  newN = newN − 1;
14                  if rand < p
15                      observed(j) = population(k);
16                      j = j + 1;
17                      newn = newn − 1;
```

(continued)

(continued)

```
18                        end
19                     end
20                     maximum = max(observed);
21                     error(loop) = (((n + 1/n)*maximum−1−N)*100/N;
22                  end
23                  subplot(2,2,z)
24                  hist(error,100)
25                  xlabel('percent error')
26                  ylabel('number of simulations (10,000 total)')
27                  title_str = ['sample size = 'int2str(size(z)) '%'];
28                  title(title_str)
29            end
```

Line 21 computes the percentage error made by our estimation formula and stores it in the vector error (which has, of course, 10,000 elements). Once error is filled, lines 23 through 28 take care of printing and labeling a histogram plot of the values in error. There will be four such plots, of course, and I used MATLAB's subplot command to place all four plots on a single page to allow easy comparison. The format of the command is that subplot(m,n,z) specifies plot z out of a total of mn plots, and so subplot(2,2,z) means plot z—where $1 \leq z \leq 4$ (take another look at line 02)—out of four plots. The positioning of the plots on a page is defined as follows: plot 1 is upper left, plot 2 is upper right, plot 3 is lower left, and plot 4 is lower right.

In line 24 the actual histogram of error is generated by the incredibly useful MATLAB command hist; i.e., hist(error,100) generates a 100-bin histogram of the elements of error. Lines 25 and 26 produce the obvious x-axis and y-axis labels; in MATLAB everything between the beginning 'and the ending' is the so-called *string variable*, which is printed. Line 27 defines yet another string variable (which will be the title line of the current plot) as the *concatenation* of three substrings: *sample size*, *int2str(size(z))*, and %. The middle substring, different for each of the four plots, is created by MATLAB's integer-to-string command, using the current value of size(z) as the argument. When all four subplots have been so generated, the program terminates.

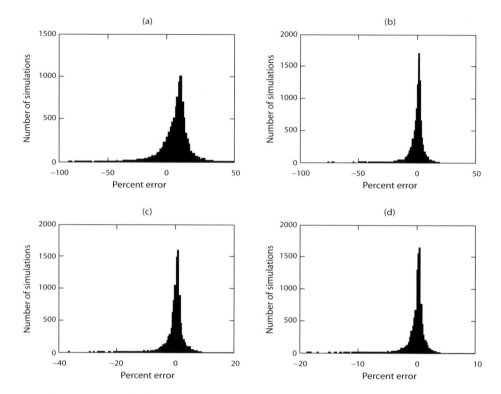

Figure S12.1. The bigger the sample size, the smaller the error (on average). (a) Sample size = 2%. (b) Sample size = 5%. (c) Sample size = 10%. (d) Sample size = 20%.

Figure S12.1 shows the results when estimate.m was run. You can see that as the sample size (percentages) increase, the spread in the estimation error decreases (which makes sense), and that once the sample size percentage has reached at least 10%, then the vast majority of the estimates are less than ±5% in error. Further, notice that independent of the sample size, the error histograms are all very nearly symmetrical around zero error. That is, the plots suggest that the average estimation error is zero, independent of the sample size. This is what statisticians mean when they say an estimation formula is unbiased.

13. *A Police Patrol Problem*

Figure S13.1 shows a logical flow diagram of the simulation of this problem, and the code patrol.m follows the flow diagram faithfully. The code assumes that all of the patrol cars are randomly patrolling the road; for scenario (a), that of a single patrol car sitting at $y = 1/2$, only a simple extra line of code is required, which I'll tell you about in just a bit. To make the code crystal clear, the following comments will almost certainly be helpful. The variables totalgrassdistance and totalconcretedistance are the total travel distances accumulated (in one million simulations) for the two traffic lanes separated by a grassy median and by a concrete barrier, respectively. With number as the number of patrol cars being used in each simulation, the row vectors gdist and cdist are defined such that gdist(k) and cdist(k) are the travel distances to the accident of the kth patrol car in the grassy median and concrete barrier cases, respectively. The value of the variable acclane is the lane in which the accident occurs, and the row vector patlane is defined to be such that the value patlane(k) is the lane which the kth patrol car occupies when notified that an accident has occurred. The row vector y is defined to be such that the value y(k) is the distance of the kth patrol car from the left end of the stretch of road when notified an accident has occurred.

For scenario (a), that of a single patrol car sitting at $y = 1/2$ (in either lane), we need only insert the line y(1)=0.5 between lines 20 and 21, and then run patrol.m with one patrol car.

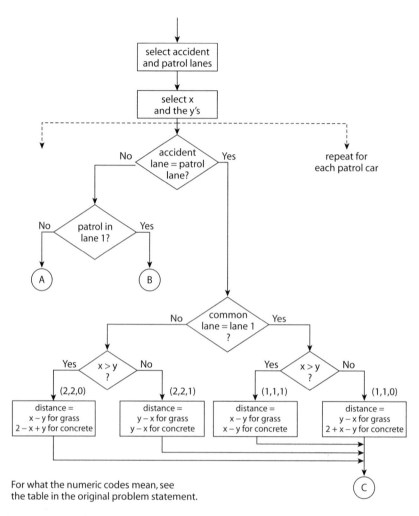

Figure S13.1. Simulating the police patrol problem. Note: For the meaning of the numeric codes, see the table in problem 13 (p. 72).

Before telling you the estimates produced by patrol.m, let me explain the theoretical results I told you about in the original problem statement, results that we can use to partially check the operation of the code. First, consider scenario (a) and (1)—that of a single patrol car sitting at $y = 1/2$ (in either lane) with a grassy median separating the traffic lanes. The distance the patrol car must drive to reach a

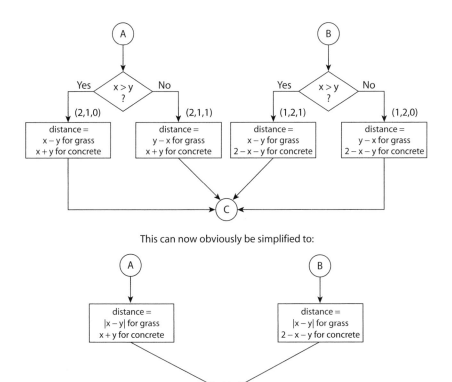

Figure S13.1. (*continued*)

random accident (in either lane) is a random variable (let's call it **Z**) given by $\mathbf{Z} = |\mathbf{X} - 1/2|$, where **X** is uniform over 0 to 1. The values of **Z** are obviously in the interval 0 to 1/2. The probability distribution function of **Z** is $F_{\mathbf{Z}}(z) = \text{Prob}(\mathbf{Z} \le z) = \text{Prob}(|\mathbf{X} - 1/2| \le z) = \text{Prob}(-z \le \mathbf{X} - 1/2 \le z) = \text{Prob}(1/2 - z \le \mathbf{X} \le 1/2 + z)$. That is, $F_{\mathbf{Z}}(z)$ is the probability that **X** is in a subinterval of length $2z$ (out of the entire interval for **X** of length 1), and since **X** is uniform, we immediately have $F_{\mathbf{Z}}(z) = 2z$, $0 \le z \le 1/2$, i.e.,

$$F_{\mathbf{Z}}(z) = \begin{cases} 2z, & 0 \le z \le \frac{1}{2}, \\ 0, & z \le 0 \\ 1, & z \le \frac{1}{2}. \end{cases}$$

patrol.m

```
01    number = input('How many patrol cars?');
02    totalgrassdistance = 0;
03    totalconcretedistance = 0;
04    for loop = 1:1000000
05        gdist = zeros(1,number);
06        cdist = zeros(1,number);
07        if rand < 0.5
08            acclane = 1;
09        else
10            acclane = 2;
11        end
12        x = rand;
13        for k = 1:number
14            y(k) = rand;
15            if rand <0.5
16                patlane(k) = 1;
17            else
18                patlane(k) = 2;
19            end
20        end
21        for k  = 1:number
22            if acclane == patlane(k)
23                if acclane == 1
24                    if x > y(k)
25                        gdistance = x − y(k);
26                        cdistance = x − y(k);
27                    else
28                        gdistance = y(k) − x;
29                        cdistance = 2 + x − y(k);
30                    end
31                else
32                    if x > y(k)
33                        gdistance = x − y(k);
34                        cdistance = 2 − x + y(k);
35                    else
```

(continued)

(continued)

```
36                          gdistance = y(k) − x;
37                          cdistance = y(k) − x;
38                  end
39              end
40          else
41              if patlane(k) == 1
42                  gdistance = abs(x − y(k));
43                  cdistance = 2 − x − y(k);
44              else
45                  gdistance = abs(x − y(k));
46                  cdistance = x + y(k);
47              end
48          end
49          gdist(k) = gdistance;
50          cdist(k) = cdistance;
51      end
52      mingdistance = min(gdist);
53      mincdistance = min(cdist);
54      totalgrassdistance = totalgrassdistance + mingdistance;
55      totalconcretedistance = totalconcretedistance
             + mincdistance;
56  end
57  totalgrassdistance/1000000
58  totalconcretedistance/1000000
```

The probability density function of $\mathbf{Z}$ is

$$f_{\mathbf{Z}}(z) = \frac{d}{dz} F_{\mathbf{Z}}(z) = \begin{cases} 2, & 0 \le z \le \frac{1}{2} \\ 0, & \text{otherwise,} \end{cases}$$

and so the average (or expected) value of $\mathbf{Z}$ is $E(\mathbf{Z})$, where

$$E(\mathbf{Z}) = \int_0^{\frac{1}{2}} z f_{\mathbf{Z}}(z)\, dz = \int_0^{\frac{1}{2}} 2z\, dz = (z^2 \mid_0^{1/2} = \frac{1}{4},$$

as claimed in the original problem statement.

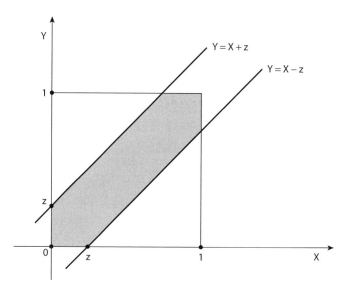

Figure S13.2. Calculating the distribution function for scenario (B) and (1).

If a single patrol car is not sitting at $y = 1/2$ waiting for a radio call to respond to an accident but rather is located at a random location (in either lane), then we have scenario (b) and (1), and the random variable $\mathbf{Z}$ is now given by $\mathbf{Z} = |\mathbf{X} - \mathbf{Y}|$, where both $\mathbf{X}$ and $\mathbf{Y}$ are independently uniform from 0 to 1. The possible values of $\mathbf{Z}$ are now in the interval 0 to 1. The distribution function of $\mathbf{Z}$ is now given by

$$F_{\mathbf{Z}}(z) = \text{Prob}(\mathbf{Z} \leq z) = \text{Prob}(|\mathbf{X} - \mathbf{Y}| \leq z) = \text{Prob}(-z \leq \mathbf{X} - \mathbf{Y} \leq z)$$

$$= \text{Prob}(-z - \mathbf{X} \leq -\mathbf{Y} \leq z - \mathbf{X}) = \text{Prob}(z + \mathbf{X} \geq \mathbf{Y} \geq -z + \mathbf{X})$$

or

$$F_{\mathbf{Z}}(z) = \text{Prob}(\mathbf{X} - z \leq \mathbf{Y} \leq \mathbf{X} + z).$$

This probability is the probability of the shaded area in Figure S13.2 (take another look at the solution to Problem 2, too), which, because $\mathbf{X}$ and $\mathbf{Y}$ are independently uniform, is geometrically given by

$$F_{\mathbf{Z}}(z) = 1 - 2\left[\tfrac{1}{2}(1 - z)^2\right] = 1 - (1 - z)^2, \quad 0 \leq z \leq 1.$$

To find $E(\mathbf{Z})$ in this case we could do as before, i.e., find $f_{\mathbf{Z}}(z)$ by differentiating $F_{\mathbf{Z}}(z)$ and then evaluating $\int_0^1 z f_{\mathbf{Z}}(z) dz$, but it is more direct to use the following easily established[1] result: $E(\mathbf{Z}) = 1 - \int_0^1 F_{\mathbf{Z}}(z) dz$. Thus,

$$E(\mathbf{Z}) = 1 - \int_0^1 \{1 - (1-z)^2\} dz = 1 - \int_0^1 dz + \int_0^1 (1-z)^2 dz$$

or

$$E(\mathbf{Z}) = \int_0^1 (1-z)^2 dz.$$

This integral is easy to do: just change variable to $u = 1 - z$, and we get

$$E(\mathbf{Z}) = \int_0^1 u^2 du = (\tfrac{1}{3} u^3 |_0^1 = \tfrac{1}{3}$$

as claimed in the original problem statement.

Finally, suppose that we have scenario (a) and (2): a single patrol car sitting at $y = 1/2$ with a concrete median. Suppose the patrol car is in lane 1. If an accident occurs at $x > 1/2$, in lane 1, then the response distance is $x - 1/2$, and that occurs with probability $1/4$. If the accident is in lane 1, with $x < 1/2$, then the response distance is $3/2 + x$, and that occurs with probability $1/4$. And if the accident is in lane 2 (for any value of x), then the response distance is $3/2 - x$ and that occurs with probability $1/2$. Thus, given that the patrol car is in lane 1, the average response distance is

$$\tfrac{1}{4}(x - \tfrac{1}{2}) + \tfrac{1}{4}(\tfrac{3}{2} + x) + \tfrac{1}{2}(\tfrac{3}{2} - x) = 1.$$

By symmetry, if it is given that the patrol car is in lane 2, then the average response distance is also 1. No matter which lane the patrol car happens to be sitting in, the average response distance (with a concrete median) is 1.

The following table shows the estimates produced by patrol.m. As you would expect, the average response distance to an accident is always more for the concrete median case, compared to the grassy median case. (Notice, too, that the code's results agree quite well with the three specific cases we calculated theoretically: scenarios (a) and (1), (b) and (1), and (a) and (2).) What the code really tells us, however, is that the average distance is considerably more for concrete compared to grass. And it tells us that while for a grassy median, the single patrol car sitting at the middle of the road is better than a single patrolling car, for a concrete median there is no difference. As a final comment on the code's estimates, the numbers in the table hint at the following analytical expressions for the average response distance, as a function of n (the number of independent, random patrol cars):

$$\frac{1}{2n+1} \qquad \text{for a grassy median}$$

and

$$\frac{2}{n+1} \qquad \text{for a concrete median.}$$

If these expressions are correct, then as n increases the ratio of the average concrete response distance to the average grassy response distance approaches 4. Can you prove (or disprove) these guesses?

	Average response distance by median	
Nature of patrol	*Grass*	*Concrete*
One car at 1/2	0.2498	1.0004
One random car	0.3333	0.9997
Two random cars	0.2085	0.6671
Three random cars	0.1500	0.5005
Four random cars	0.1167	0.3999
Nine random cars	0.0545	0.2001

References and Notes

1. We start with $E(\mathbf{Z}) = \int_0^1 z f_{\mathbf{Z}}(z)\,dz$. Then, integrating by parts, i.e., using the well-known formula from calculus $\int_0^1 u\,dv = (uv\,|_0^1) - \int_0^1 v\,du$, with $u = z$ and $dv = f_{\mathbf{Z}}(z)\,dz$, we have $v = F_{\mathbf{Z}}(z)$ and $du = dz$. Thus, $E(\mathbf{Z}) = (z F_{\mathbf{Z}}(z)\,|_0^1) - \int_0^1 F_{\mathbf{Z}}(z)\,dz = F_{\mathbf{Z}}(1) - \int_0^1 F_{\mathbf{Z}}(z)\,dz$. Since $F_{\mathbf{Z}}(1) = 1$, we have our result.

14. *Parrondo's Paradox*

The code gameb.m simulates game **B**. The logic is pretty straightforward, but in a few places the code does take elementary advantage of MATLAB's vector capabilities. After line 01 establishes the value of ϵ, line 02 creates the row vector Mtotal of length 100, with all of its elements initially equal to zero. Mtotal(k) will be, when gameb.m ends, the total capital of all 10,000 sequences at the end of k games, $1 \leq k \leq 100$. (Line 26 divides Mtotal by 10,000 at the completion of gameb.m to finish the calculation of the ensemble average.) Line 03 controls the 10,000 simulations gameb.m will perform, each of length 100 games. For each such simulation, line 04 initializes the starting capital m at zero. A vector analogous to Mtotal but which applies to a single 100-game sequence is samplefunction, which line 05 creates as a row vector of length 100 (with each of its elements initially set equal to zero). That is, samplefunction(k) is the capital of the gambler at the end of the kth game (kth coin flip, $1 \leq k \leq 100$) during each individual 100-game sequence.

Lines 06 through 23 play the sequences of 100 games. The only line in that set of commands that probably needs explanation is line 07, which is MATLAB's "remainder after division" command, rem; executing rem(m,3) returns the remainder after dividing m by 3, and so if m is a multiple of 3, then rem(m,3) $= 0$, which is precisely what line 07 checks for. Lines 08 through 21 simulate the coin flip for an

gameb.m
```
01      epsilon = 0.005;
02      Mtotal = zeros(1,100);
03      for loop = 1:10000
04          m = 0;
05          samplefunction = zeros(1,100);
06          for flips = 1:100
07              if rem(m,3) == 0
08                  outcome = rand;
09                  if outcome < 0.1 − epsilon
10                      m = m + 1;
11                  else
12                      m = m − 1;
13                  end
14              else
15                  outcome = rand;
16                  if outcome < 0.75 − epsilon
17                      m = m + 1;
18                  else
19                      m = m − 1;
20                  end
21              end
22              samplefunction(flips) = m;
23          end
24          Mtotal = Mtotal + samplefunction;
25      end
26      Mtotal = Mtotal/10000
```

individual game, and which sublist of commands (lines 08 through 13, or lines 15 through 20) is performed depends, of course, on which coin is selected to be flipped, i.e., on what happened in line 07. After the coin flip, line 22 records the updated capital of the gambler in the samplefunction vector. Once the for/end loop of lines 06 through 23 finishes the current sequence of 100 games, line 24 updates the

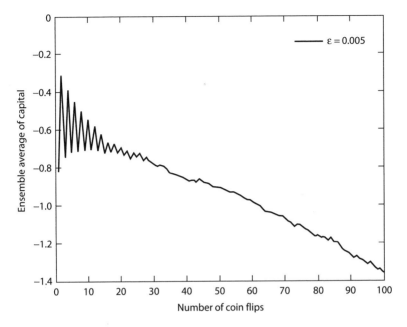

Figure S14.1. Game **B** is a losing game.

Mtotal vector, and another 100-game sequence is started. And finally, after 10,000 such sequences, line 26 computes the ensemble average as mentioned above. It is that result, Mtotal/10000, that is plotted in Figure S14.1 (I have omitted the axis labeling and plotting commands in the code for gameb.m). As that figure clearly shows, the gambler's capital oscillates for a while but, after about twenty five games or so, the oscillations have pretty much damped out, and thereafter the capital steadily becomes ever more negative. That is, game **B** is a losing game.

The code aandb.m simulates a gambler switching back and forth at random between game **A** and game **B**. That code is simply the code of gameb.m combined with additional code that implements the simple rule of game **A**. Lines 01 through 06 of aandb.m are identical to those of gameb.m, and it is not until line 07 that the decision is made on which game to play: if the variable whichgame is less than 0.5, game **A** is played by the commands in lines 09 through 14. If

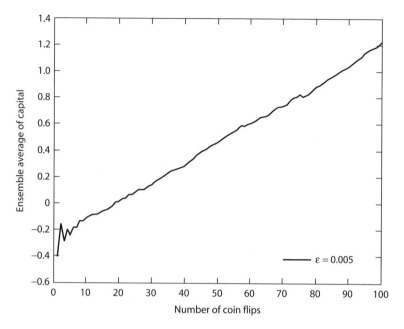

Figure S14.2. Random back-and-forth switching between games *A* and *B* *wins!*.

whichgame is greater than 0.5, game *B* is played, and, as you can verify, lines 16 through 30 of aandb.m are identical to lines 07 through 21 of gameb.m (line 31 of aandb.m is the end that completes the new "which game" loop in aandb.m started by the if in line 08). Then, to finish aandb.m, its lines 32 through 36 are identical to lines 22 through 26 of gameb.m.

When the Mtotal/10000 vector for aandb.m is plotted, the result is Figure S14.2, which clearly shows that, despite our intuitions, the capital of the gambler is (after some preliminary oscillations) steadily *increasing*! That is, randomly switching back and forth between two loser games is a winning strategy—and if you're not amazed by that, well, I find that even more amazing! Now, let me end by telling you that there is an explanation for this seemingly fantastic, counterintuitive phenomenon, and you can read all about it in a paper[1] by the man responsible for confounding (most of[2]) us.

aandb.m

```
01    epsilon = 0.005;
02    Mtotal = zeros(1,100);
03    for loop = 1:10000
04        m = 0;
05        samplefunction = zeros(1,100);
06        for flips = 1:100
07            whichgame = rand;
08            if whichgame < 0.5
09                outcome = rand;
10                if outcome < 0.5 − epsilon
11                    m = m + 1;
12                else
13                    m = m − 1;
14                end
15            else
16                if rem(m,3) == 0
17                    outcome = rand;
18                    if outcome < 0.1 − epsilon
19                        m = m + 1;
20                    else
21                        m = m − 1;
22                    end
23                else
24                    outcome = rand;
25                    if outcome < 0.75 − epsilon
26                        m = m + 1;
27                    else
28                        m = m − 1;
29                    end
30                end
31            end
32            samplefunction(flips) = m;
33        end
34        Mtotal = Mtotal + samplefunction;
35    end
36    Mtotal = Mtotal/10000
```

References and Notes

1. J.M.R. Parrondo and Luis Dinís, "Brownian Motion and Gambling: From ratchets to Paradoxical Games" (*Contemporary Physics*, March–April 2004, pp. 147–157).

2. Not everybody finds the Parrondo paradox surprising: See Ora E. Percus and Jerome K. Percus, "Can Two Wrongs Make a Right? Coin-Tossing Games and Parrondo's Paradox" (*The Mathematical Intelligencer*, Summer 2002, pp. 68–72). The authors give a good mathematical explanation of what is going on, and show that the paradox has an interesting, symmetrical twist to it: "one can win at two losing games by switching between them, but one can also lose by switching between two winning games." That is a "paradoxical" result, too, I think. And finally, much physical insight for what is going on can be gained from a paper by Hal Martin and Hans Christian von Baeyer, "Simple Games to Illustrate Parrondo's Paradox" (*American Journal of Physics*, May 2004, pp. 710–714).

15. How Long Is the Wait to Get the
Potato Salad?

To simulate the operation of the deli counter, the sequence of decisions that the clerk(s) go through, at each moment of time, is as follows, where, to simulate the flow of time, the code deli.m defines the variable clock whose value at the opening of the deli counter is zero and is 36,000 at closing; i.e., clock advances the code through simulated time in one-second increments:

(a) continue the service of the customer(s) presently being processed;

then

(b) if there is a/are clerk/s available and the queue is not empty, start the service of the person at the front of the queue (and then advance all customers left remaining in the queue);

then

(c) if a new customer has arrived and if a clerk is available, start the service of that new customer, but if a new customer has arrived and a clerk is not available add that new customer to the back of the queue;

then

> (d)   if there is still at this point a clerk available, increase that clerk's idle time;

then

> (e)   increase the waiting time for all in the queue and check to see if any customers presently being served have finished being processed.

With the completion of (e), increase clock by one (second), return to (a), and continue until the deli counter closes.

To help you understand the operation of deli.m, the following variables are defined:

(1) clerkbusy1 and clerkbusy2 are set equal to 0 if the associated clerk is available and to one if that clerk is busy; deli.m is written with the assumption that there are two clerks, and I'll tell you later how the one-clerk case is handled as a special case by forcing clerk two to never be available;

(2) the value of queuelength is the present number of people in the queue; e.g., if queuelength = 0, then the queue is empty;

(3) servicetimeremaining1 and servicetimeremaining2 are equal to the number of seconds left to complete the service of their present customers by clerk 1 and by clerk 2, respectively;

(4) queue is a 2×200 matrix, where the values of queue(1,j) and queue(2,j) are, respectively, the number of seconds the jth person in the queue has been in the queue and the (random) service time the jth person in the queue will require once he/she reaches the deli counter;

(5) the values of clerkidletime1 and clerkidletime2 are the number of seconds that clerk 1 and clerk 2 have not been involved with servicing a customer, respectively, i.e., the number of seconds of idle time for clerk 1 and clerk 2;

(6) the value of closedeli is the expected number of customers that will appear at the deli in a thirty-hour period, and so closedeli is a very conservative overestimate of the number of customers the clerk(s) will have to serve in the ten hours that the deli is open for business each day;

(7) customerarrivaltime is a vector of length closedeli, where customer-arrivaltime(j) is the arrival time of the jth customer to appear at the deli;

(8) the value of totaltime is the present sum of the total waiting times of all the customers who have so far appeared at the deli, where the waiting time for a customer is defined to be the sum of his/her time in queue (if any) and the service time by a clerk;

(9) the value of maxtotaltime is the largest, to date, total waiting time experienced by any individual customer who has been processed by a clerk;

(10) the value of totalqueuelength is the sum of the lengths of the queue, as measured each second;

(11) the value of maxqueuelength is, at any instant of time, the longest the queue has been;

(12) the value of newcustomer is such that, at any time, if newcustomer $= j$ then $j - 1$ customers have already arrived, and the code is looking for the arrival of the jth customer (newcustomer is initialized to the value 1);

(13) clock is the code's time keeper, increased by one once each second of simulated time.

Now, let's do a walkthrough of the code deli.m.

Lines 001 and 002 bring in the values of λ and μ (the variables lambda and mu). Lines 003 through 008 initialize to zero the named variables in those lines. Line 009 sets the variable closedeli to the value discussed above in item (6). Line 010, using the theoretical analysis done in Appendix 8, calculates the arrival time (in seconds, which is why the factor of 3600 is present) of the first customer, where you'll notice that the use of the ceil command (MATLAB's "round upward" command) ensures both that the result is an integer and that it is at least 1. Then, lines 011 through 013 calculate the arrival times for the next customer by first calculating the interarrival time and then adding that value onto the previous arrival time; this is done until all 30*lambda arrival times have been calculated. (Recall that three dots at the end of a line—see lines 012, 052, and 065—mean *line continuation.*) Notice that the ceil command ensures that it will never happen that two or more customers have the same arrival time (which is one of the

deli.m

```
001    lambda = input('What is the average customer arrival rate
       per hour?');
002    mu = input('What is the average customer service rate per
       hour?');
003    clock = 0;
004    clerkbusy1 = 0;clerkbusy2 = 0;
005    queuelength = 0;
006    servicetimeremaining1 = 0;servicetimeremaining2 = 0;
007    queue = zeros(2,200);
008    clerkidletime1 = 0;clerkidletime2 = 0;
009    closedeli = ceil(lambda*30);
010    customerarrivaltime(1) = ceil( - 3600*log(rand)/lambda);
011    for i = 2:closedeli
012        customerarrivaltime(i) = customerarrivaltime(i - 1)...
               + ceil( - 3600*log(rand)/lambda);
013    end
014    for i = 1:closedeli
015        customerservicetime(i) = ceil( - 3600*log(rand)/mu);
016    end
017    totaltime = 0;
018    maxtotaltime = 0;
019    totalqueuelength = 0;
020    maxqueuelength = 0;
021    newcustomer = 1;
022    clerkcount = input('Are there 1 or 2 clerks?');
023    if clerkcount == 1
024        clerkbusy2 = 1;
025        servicetimeremaining2 = 10^10;
026    end
027    while clock < 36000
028        if servicetimeremaining1 > 0
029            servicetimeremaining1 = servicetimeremaining1 - 1;
030        end
031        if servicetimeremaining2 > 0
032            servicetimeremaining2 = servicetimeremaining2 - 1;
```

(continued)

(continued)

```
033        end
034        if (clerkbusy1 == 0 | clerkbusy2 == 0)&(queuelength > 0)
035            if clerkbusy1 == 0
036                clerkbusy1 = 1;
037                servicetimeremaining1 = queue(2,1);
038            else
039                clerkbusy2 = 1;
040                servicetimeremaining2 = queue(2,1);
041            end
042            totaltime = totaltime + queue(1,1) + queue(2,1);
043            if queue(1,1) + queue(2,1) > maxtotaltime
044                maxtotaltime = queue(1,1) + queue(2,1);
045            end
046            for i = 1:queuelength
047                queue(1,i) = queue(1,i + 1);
048                queue(2,i) = queue(2,i + 1);
049            end
050            queuelength = queuelength − 1;
051        end
052        if (clock == customerarrivaltime(newcustomer))&...
                    (clerkbusy1 == 0 | clerkbusy2 == 0)
053            if clerkbusy1 == 0
054                clerkbusy1 = 1;
055                servicetimeremaining1 = customerservicetime
                        (newcustomer);
056            else
057                clerkbusy2 = 1;
058                servicetimeremaining2 = customerservicetime
                        (newcustomer);
059            end
060            totaltime = totaltime + customerservicetime
                    (newcustomer);
061            if customerservicetime(newcustomer)
                    > maxtotaltime;
062                maxtotaltime = customerservicetime
                        (newcustomer);
```

(continued)

(continued)

```
063              end
064              newcustomer = newcustomer + 1;
065          elseif (clock == customerarrivaltime(newcustomer))&...
                 (clerkbusy1 == 1&clerkbusy2 == 1)
066              queuelength = queuelength + 1;
067              queue(1,queuelength) = 1;
068              queue(2,queuelength) = customerservicetime
                     (newcustomer);
069              newcustomer = newcustomer + 1;
070          end
071          if clerkbusy1 == 0
072              clerkidletime1 = clerkidletime1 + 1;
073          end
074          if clerkbusy2 == 0
075              clerkidletime2 = clerkidletime2 + 1;
076          end
077          for i = 1:queuelength
078              queue(1,i) = queue(1,i) + 1;
079          end
080          if servicetimeremaining1 == 0
081              clerkbusy1 = 0;
082          end
083          if servicetimeremaining2 == 0
084              clerkbusy2 = 0;
085          end
086          totalqueuelength = totalqueuelength + queuelength;
087          if queuelength > maxqueuelength
088              maxqueuelength = queuelength;
089          end
090          clock = clock + 1;
091      end
092      disp('average total time at deli = '),disp(totaltime/
             (newcustomer − 1 − queuelength))
093      disp('maximum time at deli  = '),disp(maxtotaltime)
```

(continued)

(continued)

094   disp('average length of queue = '),disp(totalqueuelength/
          clock)
095   disp('maximum length of queue = '), disp(maxqueuelength)
096   disp('percent idle time for clerk 1='),disp(100*clerkidletime1/
          clock)
097   disp('percent idle time for clerk 2='), disp(100*clerkidletime2/
          clock)

---

fundamental assumptions of a Poisson process). Lines 014 through 016 calculate the service times (in seconds) that each of those customers will require once they reach an available clerk. Lines 017 through 021 initialize the named variables in those lines. Lines 022 through 026 determine if we are doing a one-clerk or a two-clerk simulation; if it is two clerks then nothing is actually done in those lines. But if it is one clerk, then clerkbusy2 is set equal to 1 (i.e., clerk 2 is busy) and servicetimeremaining2 is set equal to 10^{10} seconds (i.e., clerk 2 will *always* be busy).

Lines 027 and 091 define the outer while loop that runs deli.m through one ten-hour day. Lines 028 through 033 implement task (a). Lines 034 through 051 implement task (b); in particular, lines 046 through 051 advance the queue. Lines 052 through 070 implement task (c); in particular, lines 066 through 069 add a new customer to the back of the queue (line 067 initializes the time in queue for that customer at one second). Lines 071 through 076 implement task (d). Lines 077 through 085 implement task (e). Lines 086 through 089 update the variables totalqueuelength and maxqueuelength, and finally, clock is incremented in line 090. Then the whole business is repeated 35,999 more times. When the while loop is finally exited, lines 092 through 097 give us the code's estimates to the answers we are after — the disp command is MATLAB's 'display,' i.e., screenprint, command. The logic behind line 092, in particular, is that the average total time at the deli is totaltime divided by the number of customers who have been served, i.e., by customers-1 (which equals the total number of customers who have arrived at the deli) minus the number of customers who are in the queue waiting for service when the deli closes.

The following two tables are the output generated by deli.m for the two cases of $\lambda = 30$ and $\mu = 40$, and $\lambda = 30$ and $\mu = 25$, five times each for one and then for two clerks. (Each of these individual simulations, on a three-GHz machine with one GByte of RAM, took between five and seven seconds.) There is, of course, variation from simulation to simulation, but not more than one might reasonably expect in real life, day to day. When $\lambda = 30$ and $\mu = 40$, we have a case where a clerk can process customers, on average, faster than they arrive, and so with a single clerk we see he is idle about 18% to 29% of the time. Despite that, while the average queue length isn't worrisome (1 to 3), the maximum queue length *is* of concern (9 to 13). Also, while the average total time for a customer at the deli isn't terribly long (four to seven minutes), the unluckiest of customers experience total times three to four times longer! This is where deli.m proves its value, then, in telling us what we gain by adding a second clerk: the average total time is cut by a factor of about three, as is the maximum total time, and the maximum queue length is reduced by nearly a factor of two. The price paid for these improvements in service is that the idle time for the original clerk is doubled, and the idle time of the new clerk is even higher. Whether or not the second clerk is worth it is a management

$\lambda = 30$ and $\mu = 40$

Average total time (sec.)	Maximum total time (sec.)	Average queue length	Maximum queue length	Clerk 1 idle time (%)	Clerk 2 idle time (%)
One clerk					
426	1,717	2.8	13	18.4	×
347	1,275	2.3	10	20.1	×
244	1,028	1.3	11	28.6	×
294	866	1.5	9	27.8	×
319	1,012	2.1	10	18.8	×
Two clerks					
96	632	0.09	4	57.4	75.7
128	600	0.23	8	48.7	67.5
102	348	0.12	4	55.9	71.2
102	434	0.08	3	53.7	76.1
109	578	0.14	3	46.2	62.3

$\lambda = 30$ and $\mu = 25$

Average total time (sec.)	Maximum total time (sec.)	Average queue length	Maximum queue length	Clerk 1 idle time (%)	Clerk 2 idle time (%)
One clerk					
3,535	9,064	29.8	70	1.14	×
1,785	6,491	15.5	46	6.7	×
1,958	4,584	13.4	32	6	×
4,789	7,549	33.4	53	0.33	×
3,118	7,511	26.1	51	0.29	×
Two clerks					
199	758	0.5	7	36.3	51.3
291	1,336	1.2	9	27.2	41.8
260	1,292	1	9	23.6	32.3
172	955	0.4	5	41.9	50.4
182	1,042	0.3	5	38.8	53.4

decision, but at least now deli.m has given management some numbers to consider.

The impact of a second clerk is even more dramatic in the case where customers arrive faster, on average, than a single clerk can process them ($\lambda = 30$ and $\mu = 25$).

16. The Appeals Court Paradox

The code jury.m simulates our five-judge appeals court. Line 01 sets the value of $p(k)$ equal to the probability the kth judge makes a correct decision, where A is judge 1, B is judge 2, and so on. Line 02 sets the variable mistakes equal to zero; at the end of ten million simulated deliberations its value will be the number of incorrect court decisions. Lines 03 and 16 define the for/end loop that executes the ten million deliberations. At the start of each deliberation line 04 sets the variable majority to zero; majority will be set equal to one if this deliberation results in three or more incorrect votes (in line 13). Line 05 sets the five-element row vector votes to zero, where votes(k) will be set equal to one if the kth judge casts an incorrect vote. Lines 06 through 10 determine the votes for each of the judges; in the if/end loop of lines 07 through 09, with probability $1 - p(k)$ the vote of the kth judge is an incorrect vote. Lines 11 through 14 determine the number of incorrect votes and, if that sum exceeds two, then majority is set equal to one, which indicates that the court has made an incorrect decision. In line 15 mistakes is incremented by majority (that is, by zero if no mistake has been made, or by one if a mistake was made). Finally, line 17 gives us the code's estimate of the probability that the court delivers an erroneous decision. To partially check the code, if all the $p(k)$ are set equal to zero or to one (all the judges are either always wrong or always correct), then the court has a probability of making a mistake of either 1 or 0, respectively.

```
jury.m
01    p(1) = 0.95;p(2) = 0.95;p(3) = 0.9;p(4) = 0.9;p(5) = 0.8;
02    mistakes = 0;
03    for loop = 1:10000000
04        majority = 0;
05        votes = zeros(1,5);
06        for k = 1:5
07            if rand > p(k)
08                votes(k) = 1;
09            end
10        end
11        result = sum(votes);
12        if result > 2
13            majority = 1;
14        end
15        mistakes = mistakes + majority;
16    end
17    mistakes/10000000
```

When run with the probabilities given in the original problem statement (see line 01), jury.m produced a probability of 0.0070419, or about a 0.7% chance that the court makes a mistake. To see what happens when judge E no longer votes independently but rather always votes as does judge A, it is only necessary to insert one additional line (between lines 10 and 11): votes(5)=votes(1); this forces E's vote to match A's. This seemingly innocent change results in jury.m producing an estimate of the probability the court errs of 0.0119615, or about 1.2%. Thus, if the worst judge follows the lead of the best judge, then we have an increased (almost doubled) probability that the court errs! What happens to the concept of setting a good example? Are you surprised by this result? I certainly am, and in fact most people are.[1]

References and Notes

1. This problem was inspired by the discussion in a book by the Hungarian mathematician Gábor J. Székely, with the fascinating title *Paradoxes in*

Probability Theory and Mathematical Statistics (D. Reidel, 1986, p. 171). The very last line in Professor Székely's excellent book expresses perfectly my own philosophy about probability: "Probability theory has evolved as a symbolic counterpart of the random universe [and] it is to be hoped that the paradoxes in this book will help the reader to find the best way through our random world."

17. Waiting for Buses

Our randomly arriving bus rider must obviously arrive at the bus stop between some two consecutive hours, and we lose no generality by labeling those two hours as hour 0 and hour 1. The given theoretical value of one-half hour for the average waiting time until a bus arrives, in the $n = 1$ case, immediately follows from the observation that the possible waiting times vary from 0 to 1 (hour), and the average of such a random quantity (with a uniform distribution) is $1/2$. We can compute the theoretical value for the $n = 2$ case with only slightly more difficulty from the knowledge that the hour-on-the-hour bus arrives at time 0 and the second bus line's bus arrives at time x (where x is uniformly distributed from 0 to 1). Here's how.

A randomly arriving rider has probability x of arriving at the stop between time 0 and time x (and so has an average waiting time of $1/2x$), and similarly that rider has probability $1 - x$ of arriving at the stop between time x and time 1 (and so has an average waiting time of $1/2(1 - x)$). So, if we denote the waiting time by $\mathbf{W}$, we have

$$\mathbf{W} = \begin{cases} \dfrac{1}{2}x & \text{with probability } x \\[2mm] \dfrac{1}{2}(1 - x) & \text{with probability } 1 - x. \end{cases}$$

The conditional expected value of **W** is thus

$$E[\mathbf{W} \mid x] = \left(\frac{1}{2}x\right)(x) + \left[\frac{1}{2}(1-x)\right](1-x) = \frac{1}{2}x^2 + \frac{1}{2}(1-x)^2$$

$$= \frac{1}{2}x^2 + \frac{1}{2} - x + \frac{1}{2}x^2$$

$$= x^2 - x + \frac{1}{2}.$$

Now, since $E[\mathbf{W} \mid x]$ is a conditional expectation (i.e., conditioned on x), to find the expected waiting time $E[\mathbf{W}]$, we need to evaluate the integral

$$E[\mathbf{W}] = \int_{-\infty}^{\infty} E[\mathbf{W} \mid x] f_{\mathbf{X}}(x)\, dx,$$

where $f_{\mathbf{X}}(x)$ is the probability density function of the random variable **X** (the random variable describing the offset of the second bus line's arrival from the on-the-hour bus line's arrival). Since we are assuming that **X** is uniform from 0 to 1, we have

$$E[\mathbf{W}] = \int_{0}^{1} \left(x^2 - x + \frac{1}{2}\right) dx = \left(\frac{1}{3}x^3 - \frac{1}{2}x^2 + \frac{1}{2}x\right)\Big|_{0}^{1} = \frac{1}{3} - \frac{1}{2} + \frac{1}{2} = \frac{1}{3},$$

as given in the original statement of the problem.

To work out the theoretical average waiting time for n independently scheduled bus lines is a bit more work, as it involves evaluating an $(n-1)$-dimensional integral! I'll not tell you the theoretical answer until after we've taken a look at the simulation results produced by the code bus.m. Line 01 allows for simulating any number of bus lines (the variable n), and line 02 initializes the variable totalwaitingtime to zero (this variable's value will be the total waiting time experienced by one million riders who randomly arrive at the bus stop at any time between time 0 and time 1). Lines 03 and 25 define the for/end loop that cycles bus.m through its one million simulations. Line 04, the start of each individual simulation, initializes the vector busarrivals (of length n) to all zeros. Its first element, busarrivals(1), is indeed zero, as that is the arrival time of the hour-on-the-hour bus. The remaining $n-1$ elements are randomly generated by lines 05 through 07;

busarrivals will be used to generate the arrival times of the n buses after some final processing by line 09 (line 08 sets the arrival time of the rider—the variable riderarrival—to a random time between 0 and 1). Line 09 sorts the vector busarrivals into ascending order, i.e., after line 09 is executed we have the vector sortedbus, where

$$0 < \text{sortedbus(1)} = 0 \leq \text{sortedbus(2)} \leq \cdots \leq \text{sortedbus(n)} \leq 1.$$

Lines 10 through 23 form the for/end loop that operates on sortedbus to determine the waiting time until the first available bus. Lines 10 and 11 simply say that if the rider's arrival time is *after* the last bus has arrived (and departed), then the rider must wait until the hour-on-the-hour bus arrives (at time 1). If that condition fails, however, then there is an earlier bus that the rider can catch, i.e., the first bus to arrive after the rider's arrival time. The logic that determines which bus that is is contained in lines 13 through 22. The variable test is initially set equal to 1—the value of test will control the termination of the while loop of lines 15 through 22. Starting with sortedbus(2), the code asks if the rider arrived *before* that bus—if so, then the rider's waiting time is computed in line 17 and test is set equal to 0 (to terminate the while loop). If not so, however, the index into sortedbus is incremented by one (line 20), and the question is then asked again. This process is certain at some point to assign a value to the variable waitingtime. Line 24 uses waitingtime to update totalwaitingtime, and then another simulation is performed. Line 26 calculates the average of one million waiting times.

When I ran bus.m for $1 \leq n \leq 5$, the code produced the following results:

n	*Average waiting time (hours)*
1	0.4996
2	0.3335
3	0.2503
4	0.2001
5	0.1667

From these numbers I think it is an obvious guess that the average waiting time (in hours) for n bus lines is given by $\frac{1}{n+1}$, a guess that can be analytically confirmed.[1]

---

bus.m

```
01    n = input('How many bus lines?')
02    totalwaitingtime = 0;
03    for loop = 1:1000000
04        busarrivals = zeros(1,n);
05        for j = 2:n
06            busarrivals(j) = rand;
07        end
08        riderarrival = rand;
09        sortedbus = sort(busarrivals);
10        if riderarrival > sortedbus(n)
11            waitingtime = 1 − riderarrival;
12        else
13            test = 1;
14            j = 2;
15            while test == 1
16                if riderarrival < sortedbus(j)
17                    waitingtime = sortedbus(j) − riderarrival;
18                    test = 0;
19                else
20                    j = j + 1;
21                end
22            end
23        end
24        totalwaitingtime = totalwaitingtime + waitingtime;
25    end
26    totalwaitingtime/1000000
```

References and Notes

1. Alan Sutcliff, "Waiting for a Bus" (*Mathematics Magazine*, March 1965, pp. 102–103).

18. Waiting for Stoplights

The code walk.m simulates our pedestrian's walk from $(m+1, m+1)$ to $(1,1)$ a hundred thousand times, for any given value of m. Lines 01 and 02 respectively bring in the value of m to be used and initialize the variable totalwait to zero. When the program terminates, totalwait will be the total number of red lights encountered in 100,000 simulated walks from $(m+1, m+1)$ to $(1,1)$. Lines 03 and 26 define the for/end loop that simulates a single one of those walks. At the start of a walk, lines 04 and 05 set the pedestrian at the starting point $(j,k) = (m+1, m+1)$, and line 06 initializes the variable wait to zero (this variable will, at the end of the walk, be the number of red lights encountered). Lines 07 and 13 define a while loop that executes as long as the pedestrian has not yet reached a boundary line, i.e., as long as both j and k are greater than one. As long as that is true, at each step (the walking of a block), j or k is decremented by one, with equal probability, under the control of the if/end loop in lines 08 through 12.

When the while loop terminates, it must be because a boundary line has been reached, i.e., either $j = 1$ (and $k > 1$) or $k = 1$ (and $j > 1$). From then on the walk proceeds straight in towards the destination point $(1,1)$ along that boundary line. The if/end loop in lines 14 through 18 sets the variable z equal to the value of the one of j and k that is greater than 1. Then, the while loop in lines 19 through 24 simply decrements z by one (in line 23) after the internal if/end loop in lines 20 through 22 increments wait by one with probability 1/2 (a red light

is encountered) or doesn't increment wait with probability 1/2 (a green
light is encountered). This continues until z has been counted down to
1, which means the pedestrian has at last reached (1,1), at which point
line 25 updates totalwait. A new walk is then started. At the completion
of 100,000 walks, line 27 gives us walk.m's estimate for the average
number of red lights encountered on a walk.

---

walk.m

```
01    m = input('What is m?');
02    totalwait = 0;
03    for loop = 1:100000
04        j = m + 1;
05        k = m + 1;
06        wait = 0;
07        while j > 1 & k > 1
08            if rand < 0.5
09                j = j − 1;
10            else
11                k = k − 1;
12            end
13        end
14        if j == 1
15            z = k;
16        else
17            z = j;
18        end
19        while z > 1
20            if rand < 0.5
21                wait = wait + 1;
22            end
23            z = z − 1;
24        end
25        totalwait = totalwait + wait;
26    end
27    totalwait/100000
```

Before I tell you what walk.m produces for estimates to the questions I put to you, let me show you how to solve the problem in the form of a recurrence formulation that will give precise numerical answers. That will give us checks on just how well the Monte Carlo simulation has performed. We start by defining $E(j,k)$ as the expected (i.e., average) number of red lights encountered on a walk from (j,k) to $(1,1)$. Obviously, $E(1,1) = 0$. As long as j and k are both greater than 1, i.e., as long as the pedestrian is not yet on either of the two boundary lines, then with probability $1/2$ she will move through the intersection at (j,k) in the direction that eventually takes her to the intersection at $(j - 1,k)$, and with probability $1/2$ she will move through the intersection at (j, k) in the direction that eventually takes her to the intersection at $(j,k - 1)$. We thus have the recursion

(a) $E(j,k) = \frac{1}{2}E(j - 1, k) + \frac{1}{2}E(j, k - 1), j, k > 1.$

If, however, either j or k is equal to 1, then we have the so-called boundary conditions:

(b) $E(1,k) = \frac{1}{2}(k - 1), k \geq 1$

because, once on the vertical boundary line she can only move downward along that line and, at $k - 1$ intersections from her final destination, she can expect half of them to have a red light; in the same way

(c) $E(j,1) = \frac{1}{2}(j - 1), j \geq 1.$

The answer to our problem is, by definition, $E(m + 1, m + 1)$, which we can find by using, over and over, the above three equations. For example, suppose $m = 2$. Thus, the pedestrian's walk starts at $(3, 3)$, as shown in Figure S18.1. We have, from (a),

$$E(2,2) = \frac{1}{2}E(1, 2) + \frac{1}{2}E(2, 1)$$

and since, from (b) and (c)

$$E(1,2) = \frac{1}{2} \text{ and } E(2, 1) = \frac{1}{2}$$

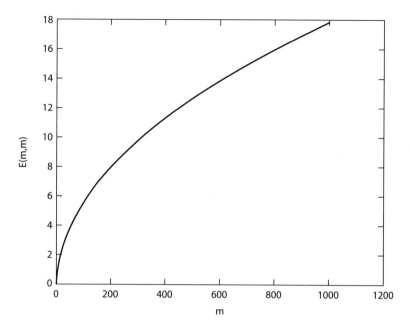

Figure S18.1. Waiting for Red Lights.

we have

$$E(2,2) = \tfrac{1}{2} \times \tfrac{1}{2} + \tfrac{1}{2} \times \tfrac{1}{2} = \tfrac{1}{2}.$$

Then,

$$E(2,3) = \tfrac{1}{2}E(1,3) + \tfrac{1}{2}E(2,2) = \tfrac{1}{2} \times 1 + \tfrac{1}{2} \times \tfrac{1}{2} = \tfrac{3}{4}$$

and

$$E(3,2) = \tfrac{1}{2}E(2,2) + \tfrac{1}{2}E(3,1) = \tfrac{1}{2} \times \tfrac{1}{2} + \tfrac{1}{2} \times 1 = \tfrac{3}{4}.$$

So, at last,

$$E(3,3) = \tfrac{1}{2}E(3,2) + \tfrac{1}{2}E(2,3) = \tfrac{1}{2} \times \tfrac{3}{4} + \tfrac{1}{2} \times \tfrac{3}{4} = \tfrac{3}{4} = 0.75.$$

We could continue with this process, working our way by hand through the one million intersections that are in the $1{,}000 \times 1{,}000$ array with its northwest corner at $(1001,1001)$, but for the obvious reason, it is much more desirable to use a computer! The code easywalk.m does that job (I've not included the labeling commands), and a plot of $E(m+1, m+1)$ for $0 \le m \le 1000$ is shown in Figure S18.1.

easywalk.m

```
01    m = input('What is m?');
02    for j = 1:m + 1
03        E(j,1) = (j − 1)/2;
04        E(1,j) = E(j,1);
05    end
06    for k = 2:m + 1
07        for j = 2:m + 1;
08            E(j,k) = (E(j − 1,k) + E(j,k − 1))/2;
09        end
10    end
11    for k = 1:m + 1
12        x(k) = k;
13        y(k) = E(k,k);
14    end
15    plot(x,y)
```

In the following table are the specific values of $E(m + 1, m + 1)$ produced by walk.m for a few particular values of m, and the values produced for the same values of m by easywalk.m. As you can see, the agreement is quite good.

m	$E(m+1,m+1)$ by easywalk.m	$E(m+1,m+1)$ by walk.m
2	0.75	0.74936
5	1.23046875	1.22906
10	1.762	1.75866
20	2.5074	2.50673
50	3.97946	3.99415
100	5.6348479	5.63352
1,000	17.839	. . . .

As a final comment, it is possible to solve the recurrence equations (a), (b), and (c) to get an analytical expression for $E(m + 1, m + 1)$, rather than using those equations directly as does easywalk.m. As the author of that solution wrote,[1] with some humor,

> Encouraged by our recent success in solving a similar (but non-homogeneous) partial difference equation of first order, we applied deductive reasoning [this is a euphemism for "educated guessing"!] and found, with the help of pencil, paper, and an enormous wastepaper basket [this is the writer admitting to his "educated guessing"!] that ... [there then follows a rather complicated expression involving summations that incorporate multiple binomial coefficients with variable parameters].

That author still had to write a computer program to do an awful lot of grubby number crunching,[2] however, and it is not clear to me that his analytical expression provides any significant computational advantage over the direct code of easywalk.m. This problem considerably predates Sagan's analysis.[3]

References and Notes

1. Hans Sagan, "On Pedestrians, City Blocks, and Traffic Lights" (*Journal of Recreational Mathematics*, 21 [no. 2], 1989, pp. 116–119).

2. Sagan's BASIC code has several uses of the GO TO statement, which fell from grace some years ago. Or at any rate, it is a favorite target of academic computer scientists. It never much bothered engineers, however, and it was invented by mathematicians, who also didn't seem to find it repulsive. GO TO's, I have to admit, can result (if overused) in computer codes so twisted and convoluted as to look like some beastly spaghetti horror from the eighth dimension. (In my younger days, I also have to admit, I wrote a few codes like that!) So maybe the academic computer scientists are right.

3. Benjamin L. Schwartz, "A Pedestrian Problem" (*Journal of Recreational Mathematics* 16 [no. 1], 1983–84, pp. 61–62). Schwartz's solutions to his own problem, by a Monte Carlo simulation (but no code is provided) and a recurrence formula, appeared in the *Journal of Recreational Mathematics* (17 [no. 1], 1984–85, pp. 73–75). The mathematical formulation of the problem by Schwartz, while correct, is in my opinion just a bit awkward, and so I've followed the formulation used by Sagan (notes 1 and 2) in my presentation here, with one exception. Sagan uses $(0,0)$ as the pedestrian's destination, while I have used $(1,1)$. The reason for this change is simply that Sagan used BASIC, which allows zero indexing into arrays—which MATLAB does not allow. This is a trivial change, of course, as a walk from $(m+1, n+1)$ to $(1,1)$ is equivalent to a walk from (m,n) to $(0,0)$.

19. Electing Emperors and Popes

The code election.m gives us estimates for the election probability of a leader for a group with N members, of whom $n \leq N$ are the dark horse candidates who receive among themselves all N notes, with at least M votes required for election. The operation of the code hinges on the row vector result, of length n; if we number the n dark horse candidates from 1 to n, then the value of result(j) will be the number of votes received by candidate j. To start, lines 01 through 05 define the basic variables of N, n, and M, as well as leader (the number of times a vote actually elects a leader after 100,000 simulations), and mayvoteforself (if set equal to zero, nobody is allowed to vote for him or herself, while if set equal to 1 one *is* allowed to vote for oneself). The for/end loop contained within lines 06 and 21 define an individual vote. Line 07 initializes result to all zeros before each vote, and then the for/end loop defined by lines 08 and 16 generates a random vote from each of the N members of the group.

Without loss of generality, we can assume that the first n members of the group are the n dark horse candidates who can be voted for, which explains line 09; the variable select is there set equal to one of the integers $1, 2, \ldots, n$. The for/end loop in lines 10 through 14 automatically accept this value for select if voting for oneself is allowed (i.e., if mayvoteforself = 1), but if voting for oneself is not allowed, then the code checks to see if the current vote has been cast by the voter for himself (is the value of the loop control variable ballot equal to the value of select?). If a voter has voted for himself, then the code cycles in the while loop of lines 11, 12, and 13, generating new values for select until the current voter does vote for someone other than himself. Line 15 updates the vector result, which, as explained above, keeps track of how many votes each of the n dark horse candidates has received.

Once all N votes have been cast, result is examined in line 17, and the variable most is assigned a value equal to the number of votes received by the best-performing candidate. Lines 18 through 20 then check to see if most is at least equal to M—if so, a leader has been elected, and leader is incremented by one. At the completion of 100,000 simulated votes, line 22 calculates the probability a leader has been elected.

```
election.m
01    N=7;
02    n=7;
03    M=4;
04    leader=0;
05    mayvoteforself=0;
06    for loop=1:100000
07         result=zeros(1,n);
08         for ballot=1:N
09              select=ceil(n*rand);
10              if mayvoteforself==0
11                   while select==ballot
12                        select=ceil(n*rand);
13                   end
14              end
15              result(select)=result(select)+1;
16         end
17         most=max(result);
18         if most> =M
19              leader=leader+1;
20         end
21    end
22    leader/100000
```

Now, before I tell you what estimates election.m produced in answer to the questions posed in the original problem statement, let me show you a few simple calculations for some very elementary cases (small values of N, M, and n) that we can use to (partially) validate the code. First, suppose we have a group of three persons ($N = 3$), voting at random among two of them ($n = 2$). We'll use $M = 2$; i.e., it takes a majority vote to be elected. This case is so elementary we can easily write down all the possibilities—what mathematicians call the *sample space points* of the problem. Since each of the three persons in the group has two choices, there are eight such possibilities. As explained in above discussion of election.m, persons 1 and 2 are the dark horse candidates here, and, if we allow voting for oneself, then the following table shows which of the eight sample points results in the election of a leader; the entries in the jth column indicate for which dark horse candidate person j votes.

Person 1	Person 2	Person 3	Leader elected?
1	1	1	Y(1)
1	1	2	Y(1)
1	2	1	Y(1)
1	2	2	Y(2)
2	1	1	Y(1)
2	1	2	Y(2)
2	2	1	Y(2)
2	2	2	Y(2)

In fact, every possible vote elects a leader! So, with mayvoteforself=1, the code should produce an "estimate" of 1 for the probability of electing a leader. It should now be obvious that mayvoteforself= 0 will result in the same probability.

Let's try something just a bit more complicated. Consider the simplest possible Imperial Election, with three electors. Now $N = 3$, $n = 3$, and $M = 2$. If we write down all possible votes (with voting for yourself allowed), we find there are twenty seven sample points (as shown in the following table), with twenty one of them resulting in the election of a leader. That is, the probability of electing a leader with mayvoteforself = 1 is 21/27=7/9 = 0.77778. On the other hand, if voting for yourself is not allowed, then the number of sample space points shrinks to just eight (the ones whose rows are marked with asterisks). Since six of those eight possibilities result in electing a leader, then the probability of electing a leader with mayvoteforself = 0 is 6/8 = 3/4 = 0.75.

Person 1	*Person 2*	*Person 3*	*Leader elected?*
1	1	1	Y(1)
1	1	2	Y(1)
1	1	3	Y(1)
1	2	1	Y(1)
1	2	2	Y(2)
1	2	3	N
1	3	1	Y(1)
1	3	2	N
1	3	3	Y(3)
2	1	1	Y(1)*
2	1	2	Y(2)*
2	1	3	N
2	2	1	Y(2)
2	2	2	Y(2)
2	2	3	Y(2)
2	3	1	N*
2	3	2	Y(2)*
2	3	3	Y(3)
3	1	1	Y(1)*
3	1	2	N*
3	1	3	Y(3)
3	2	1	N
3	2	2	Y(2)
3	2	3	Y(3)
3	3	1	Y(3)*
3	3	2	Y(3)*
3	3	3	Y(3)

When election.m was run the following estimates were produced. First, suppose that voting for yourself is allowed. Then, for our first test case ($N = 3, n = 2, M = 2$), the code's estimate of the probability of electing a leader was 1 (compare to the theoretical value of 1). For our second test case ($N = 3, n = 3, M = 2$), election.m's estimate was 0.77698 (compare to the theoretical answer of 0.77778). If voting for oneself is not allowed, the results were: code(1)/theory(1), and code(0.75019)/theory(0.75000), respectively. For the Imperial Election problem ($N = 7, n = 7, M = 4$), the probability of electing a leader if one may (may not) vote for oneself is 0.07099 (0.05989).[1] And finally, for the papal election problem of 1513 ($N = 25$ and $M = 17$) election.m gave the following estimates for the probability of randomly electing a Pope from n candidates:

n	*Voting for self allowed*	*Voting for self not allowed*
2	0.10891	0.09278
3	0.00115	0.00094
4	0.00005	0.00004

Probably not too much can be concluded from the last line other than it is very unlikely that a random vote would actually elect a pope. To be really sure about the actual probabilities for $n \geq 4$, we would need to run considerably more than 100,000 simulated votes.

Some final comments. You may recall from the introduction that I claimed one use of Monte Carlo is that of using simulation results to check theoretical calculations. It was just that sort of use that first attracted me to this problem, which I first read about in a fascinating paper by Professor Anthony Lo Bello (Department of Mathematics, Allegheny College).[2] In that paper are a number of derived probability formulas that are supposed to be the answers to questions about the probability of electing a leader as a function (in the notation used here) of N, n, and M. However, when I wrote and ran election.m, there were significant differences between its estimates and the values produced by the formulas—differences large enough, in fact, that it was clear that something was not right. After carefully reading through

Professor Lo Bello's paper I identified where I believed things went wrong in his theoretical analysis—and then wrote to him about it. Professor Lo Bello quickly replied and, with admirable openness (not all writers would do this), wrote, "Yes, I certainly blew it in that paper. I subsequently handled the problem correctly. . ." and that second paper (in *The Mathematical Gazette*) is the one I cite in the original statement of this problem. Both of Professor Lo Bello's papers provided nearly all of the historical content provided in this book on the papal election problem.

References and Notes

1. The sample space of this version of the Imperial Election problem is far too large to consider writing it out point by point, as I did in our test cases. By making only some elementary arguments, however, we can still calculate how many sample points there are associated with electing a leader (which is all we need) and so arrive at one more check on the coding of election.m. First, suppose that anyone can vote for anyone. Since each of seven people can vote for any of seven people, there are a total of 7^7 points in sample space. For a particular one of the seven to be elected leader, he must receive either exactly four votes *or* exactly five votes *or* exactly six votes *or* exactly seven votes. Consider each case in turn.

- To receive exactly seven votes, the elected person must receive all the votes, and there is only one way that can happen. So there is one sample point associated with the elected person receiving seven votes.
- To receive exactly six votes, one person must *not* vote for the elected person, and that can happen in as many ways as one can pick one person from seven, i.e., in $\binom{7}{1} = 7$ ways. That one person can vote for any of the *six* people other than the elected person. So, there are $6 \times 7 = 42$ sample points associated with the elected person receiving six votes.
- To receive exactly five votes, two people must *not* vote for the elected person, and that can happen in $\binom{7}{2} = 21$ ways. Each of those two people can vote for any of six people. So, there are $21 \times 6 \times 6 = 756$ sample points associated with the elected person receiving five votes.
- To receive exactly four votes, three people must *not* vote for the elected person, and that can happen in $\binom{7}{3} = 35$ ways. Each of those three people can vote for any of six people. So, there are $35 \times 6 \times 6 \times 6 = 7,560$ sample points associated with the elected person receiving four votes.

Thus, the probability of a particular one of the seven people to be elected leader is

$$(1 + 42 + 756 + 7,560)/7^7 = 8,359/7^7.$$

Since we are interested in the probability that someone (not just a particular one) is elected leader, the probability we are after is seven times this, i.e.,

$$7 \times 8,359/7^7 = 8,359/7^6 = 0.07105,$$

which compares nicely with the Monte Carlo estimate of 0.07099.

If none of the seven people is allowed to vote for himself, there are simple modifications to the above arguments that let us calculate the new probability of randomly electing a leader. First, observe that the only way to be elected is to receive exactly four votes *or* exactly five votes *or* exactly six votes; i.e., it is not possible to receive seven votes. Again, let's consider each case in turn for the election of a particular one of the seven.

- To receive exactly six votes, everybody else votes for the elected person, which can happen in just one way. The elected person votes for any of the other six. So, there are $1 \times 6 = 6$ sample points associated with the elected person receiving six votes.
- To receive exactly five votes, there must be one person in the other six who does not vote for the elected person, which can happen in six ways. That person can vote for any of five people (neither the elected person or himself), and the elected person can vote for any of the other six. So, there are $6 \times 5 \times 6 = 180$ sample points associated with the elected person receiving five votes.
- To receive exactly four votes, there must be two persons in the other six who do not vote for the elected person, which can happen in $\binom{6}{2} = 15$ ways. Each of those two people can vote for any of five, and the elected person can vote for any of six. So, there are $15 \times 5 \times 5 \times 6 = 2,250$ sample points associated with the elected person receiving four votes.

Thus, the probability of someone of the seven people being elected leader is

$$7 \frac{6 + 180 + 2,250}{6^7} = 7 \times \frac{2,436}{6^7} = \frac{17,052}{6^7} = 0.06091,$$

which again compares nicely with the Monte Carlo estimate of 0.05989.

2. Anthony Lo Bello, "A Papal Conclave: Testing the Plausibility of a Historical Account" (*Mathematics Magazine*, September 1982, pp. 230–233).

The Very
Best

20. *An Optimal Stopping Problem*

The central idea of this problem has been around in the mathematical world since 1875, when the English mathematician Arthur Cayley (1821–1895) proposed something similar in the form of a lottery problem.[1] That problem does not follow the exact same rules that we have in our dating game problem, and so at least one writer has concluded Cayley's problem doesn't count as a historical precedent, but I don't agree. You can make up your own mind.[2] It is generally agreed, however, that our problem became widely known in 1960 only after it appeared in Martin Gardner's famous "Mathematical Games" column in *Scientific American*.[3]

Before discussing the Monte Carlo simulation code for this problem, let me show you the theoretical answer for the proper sample lot size in the special case where we are happy only in the event that we select *the* very best person in the initial entire population. We can then use this result to partially validate the code (which will, in fact, allow the generalization to selecting somebody from the top k people, where $k \geq 1$, not just $k = 1$). I'll start by defining $r - 1$ as the sample lot size that we will use, where $r \geq 1$. That is, our strategy is to reject the first $r - 1$ dates out of hand (while remembering the best of those dates), and to accept the first person thereafter who is better than the remembered best. Our problem is to find that value of r that maximizes the probability that the person we decide to accept is the very best person in the entire initial population. Next, let's write $\phi_n(r)$ as the probability we accept the very best person, as a function of n (the size of the entire initial population) and of r (one more than the size of the sample lot). Then,

$$\phi_n(r) = \sum_{j=r}^{n} \{\text{Prob}\,(j\text{th date is the very best person})\}$$

$$\times \text{Prob}\,(j\text{th date is selected})\}.$$

Since we will assume that all of the $n!$ possible permutations of how we can sequentially date n people are equally likely, then any particular person is as likely to be the jth date as is anybody else, and so

$$\text{Prob}(j\text{th date is the very best person}) = \frac{1}{n}.$$

To calculate Prob(jth date is selected), we'll treat the $r = 1$ and the $r > 1$ cases separately. The answer for $r = 1$ is immediately clear from the same reasoning as above; i.e., $r = 1$ means the sample lot size is zero, which means that you accept the very first person you date with probability one, and so the probability that person is the very best person is $1/n$, i.e., $\phi_n(1) = 1/n$. The case of $r > 1$ requires just a slightly more subtle argument.

For you to be able to select the very best person on the jth date, that person obviously must be the jth date, an event that we have argued has probability $1/n$. Further, the best of the first $r - 1$ dates (that is, the person who sets the standard for the person you will eventually select) must actually be the best in the first $j - 1$ dates. This is so because if there was somebody else in the first $j - 1$ dates who is better than the best of the first $r - 1$ dates, then that person would by definition be a date after the first $r - 1$ dates and so would be selected before the jth date! So, imagine an external observer who can see the first $j - 1$ dates all lined up in a row for you. The best of those $j - 1$ dates must occur in the first $r - 1$ dates. Since that person is as likely to be any one of the first $j - 1$ dates, then

$$\text{Prob (you select } j\text{th person)} = \frac{r - 1}{j - 1}, r > 1.$$

Thus,

$$\phi_n(r) = \sum_{j=r}^{n} \frac{1}{n} \times \frac{r-1}{j-1} = \frac{r-1}{n} \sum_{j=r}^{n} \frac{1}{j-1}, r>1.$$

To find the optimal value of r, for a given n, we need only evaluate $\phi_n(r)$ for all possible r ($1 \le r \le n$) and pick that r that gives the largest ϕ. For small n that's not too hard to do by hand, but for $n>10$ it quickly becomes tiresome. The code stopping.m does all the grubby number crunching for us, and it is what I used to produce the table (for $n = 11$) given in the original statement of the problem. (The code starts with $r = 2$ in line 03 because for $r = 1$ we already know that $\phi_n(1) = 1/n$ for any $n \ge 1$.) The code is, I think, obvious, with perhaps just lines 10 and 11 deserving a bit of explanation. Line 10 assigns the value of the maximum probability of success (of picking the very best person) to

---

stopping.m

```
01    n = input('What is n?');
02    prob = zeros(1,5000);
03    for r = 2:n
04         s = 0;
05         for j = r:n
06              s = s + 1/(j-1);
07         end
08         prob(r) = (r-1)*s/n;
09    end
10    [maximum,index] = max(prob)
11    n/(index-1)
```

---

the variable maximum, while index is assigned the location in the row vector prob of that maximum probability. The size of the sample lot is index-1, and so line 11 computes the ratio of the initial population size to the size of the sample lot. It is interesting to see how that ratio behaves as n increases, and the following table shows that behavior.

n	n/Sample lot size	Probability of selecting the very best person
5	2.5	0.43333
10	3.33333	0.39869
20	2.85714	0.38421
50	2.77777	0.37427
100	2.7027	0.37104
200	2.73972	0.36946
500	2.71739	0.36851
1,000	2.71739	0.36819
2,000	2.71739	0.36804
5,000	2.71887	0.36794

You would have to be pretty obtuse not to suspect that our ratio is approaching $e = 2.71828\ldots$ as $n \to \infty$. That is, for large n, the optimal sample lot size is the fraction $1/e = 0.36787\ldots$ of the initial population. Further, as the rightmost column suggests, the probability of selecting the best person approaches $1/e$ as well, as $n \to \infty$. To answer the specific questions asked in the original statement of the problem, I used the simulation code optimal.m. The logic of the code, for a single simulation, is shown in Figure S20.1.

After line 01 brings in the value of n (the size of the initial dating population), line 02 asks for how far down in the dating pool you are willing to settle; e.g., top = 1 means you want the very best, top = 2 means you'll be happy with either of the best two, and so on. Line 03 defines the row vector results, where results(j) will be the total number of successful simulations (out of 10,000) when the size of the sample lot is j, i.e., when the first j dates are automatically rejected. The length of results is about 80% of the maximum possible number of sample lot sizes. Lines 04 and 34 define the for/end loop that will sequentially run the code through values of the sample lot size, from 1 to about 80% of the maximum possible. When you look at Figures S20.2 and S20.3, you'll understand why it isn't necessary to be concerned about larger sample lot sizes.

optimal.m

```
01    n = input('Size of dating pool?');
02    top = input('Number of acceptable choices?');
03    results = zeros(1,ceil(4*n/5));
04    for samplesize = 1:ceil(4*n/5)
05        check = zeros(1,samplesize);
06        success = 0;
07        for loop = 1:10000
08            order = randperm(n);
09            for j = 1:samplesize
10                check(j) = order(j);
11            end
12            bestinsample = min(check);
13            exit = 0;
14            k = samplesize + 1;
15            while exit == 0
16                if order(k) < bestinsample
17                    select = order(k);
18                    exit = 1;
19                else
20                    k = k + 1;
21                end
22                if k == n + 1
23                    select = order(n);
24                    exit = 1;
25                end
26            end
27            for k = 1:top
28                if select == k
29                    success = success + 1;
30                end
31            end
32        end
33        results(samplesize) = success;
34    end
35    results = results/10000;
36    bar(results)
```

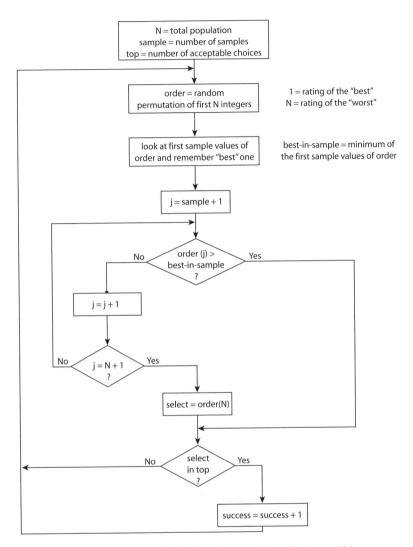

Figure S20.1. The logic of a single simulation of the dating problem.

Line 05 defines the vector check, of length samplesize, and initializes it to zero, and line 06 initializes the variable success to zero. The best person in the sample will be the minimum value in check, while the value of success will be the number of times (out of 10,000 simulations for the current value of samplesize) that the chosen one is among the top people in the initial population. Lines 07 through 32 implement the simulation logic illustrated in Figure S20.1, which is executed

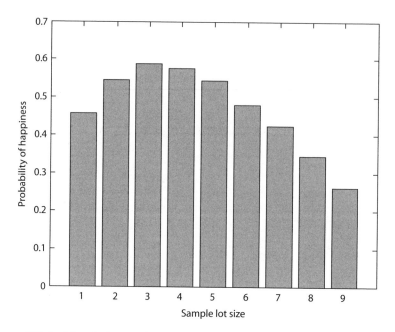

Figure S20.2. The dating problem for $n = 11$ and top $= 2$.

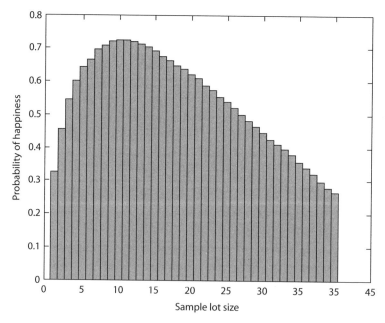

Figure S20.3. The dating problem for $n = 50$ and top $= 5$.

10,000 times for each of the values of samplesize. When the code exits that for/end loop the vector results is updated in line 33, and then a new batch of 10,000 simulations is run for the next value of samplesize. Once all the values of samplesize have been processed, line 35 normalizes results to a probability vector (the division by 10,000), and line 36 prints a bar graph of the normalized results. Figures S20.2 ($n = 11$, top $= 2$) and 20.3 ($n = 50$, top $= 5$) show what these graphs generally look like.

n	top	Optimal sample size	Probability of being happy
11	2	3	0.589
11	3	3	0.7046
11	4	2	0.7803
11	5	2	0.8365
50	1	20	0.3767
50	2	16	0.5327
50	3	13	0.6225
50	4	13	0.6851
50	5	11	0.7295

These figures tell us that the curves of the probability of happiness versus the size of the sample lot have broad maximums, which means that the value of the optimal sample lot size is not much more effective than are its near neighbors. The bar graphs rise monotonically to their maximums and remain near those maximums over an interval of values for the sample lot size, and then monotonically fall. And that's why it is not necessary to run optimal.m for all possible values of the sample lot size (I somewhat arbitrarily selected the value of ≈80% of the maximum possible value). The following table shows the estimates produced by optimal.m for all the questions asked in the original problem statement.

References and Notes

1. Leo Moser, "On a Problem of Cayley" (*Scripta Mathematica* 22, 1956, pp. 289–292).

2. Thomas S. Ferguson, "Who Solved the Secretary Problem?" (*Statistical Science*, August 1989, pp. 282–289). Ferguson's title comes from casting the problem in the form of selecting the best secretary from an initial population rather than selecting the best spouse.

3. *Scientific American* (February 1960, pp. 152–153, and March 1960, pp. 180–181).

21. *Chain Reactions, Branching Processes, and Baby Boys*

Before getting to the writing of a simulation code to estimate the answers to the questions posed in the original problem statement, let me show you a beautiful theoretical solution. I'll then use this solution to validate the code. If we have $\{p_i\}$ as the sequence of probabilities that a man will have i sons, $i \geq 1$, then define the so-called *generating function* of this sequence to be

$$f(x) = p_0 + p_1 x + p_2 x^2 + p_3 x^3 + \cdots$$

where there is no physical significance to x. What Hawkins and Ulam showed[1] is that if we calculate the iterated functions $f_n(x) = f\{f_{n-1}(x)\}$, where $f_1(x) = f(x)$, then the probability that the nth generation descendents of a lone man (the only member of the 0th generation) includes k males through a direct chain of male ancestors is the coefficient of the x^k term of the power series form of $f_n(x)$.

For our problem, using the $\{p_i\}$ sequence in the original problem statement gives

$$f(x) = 0.4825 + 0.2126x + 0.2126(0.5893)x^2 + 0.2126(0.5893)^2 x^3 + \cdots.$$

This is of the form (with $a = 0.4825$, $b = 0.2126$, and $c = 0.5893$)

$$f(x) = a + bx + bcx^2 + bc^2 x^3 + \cdots,$$

which is easily summed (because you'll notice that, except for the first term, we have a geometric series) to give

$$f(x) = \frac{a - (ac - b)x}{1 - cx} = \frac{0.4825 - 0.0717x}{1 - 0.5893x}.$$

Thus,

$$f_2(x) = \frac{0.4825 - 0.0717 \left\{ \dfrac{0.4825 - 0.0717x}{1 - 0.5893x} \right\}}{1 - 0.5893 \left\{ \dfrac{0.4825 - 0.0717x}{1 - 0.5893x} \right\}}$$

$$= \frac{0.4825 - 0.4825(0.5893x) - (0.0717)(0.4825) + 0.0717(0.0717x)}{1 - 0.5893x - (0.5893)(0.4825) + (0.5893)(0.0717x)}$$

or,

$$f_2(x) = \frac{0.4479 - 0.2792x}{0.7157 - 0.5470x}.$$

This is of the general form

$$f_2(x) = \frac{d - ex}{g - hx},$$

which by direct long division can be written as

$$f_2(x) = \frac{d}{g} + \frac{hd - eg}{g^2}x + \frac{(hd - eg)h}{g^3}x^2 + \frac{(hd - eg)h^2}{g^4}x^3 + \cdots$$

or, since $hd - eg = (0.5470)(0.4479) - (0.2792)(0.7157) = 0.0452$, we have

$$f_2(x) = \frac{0.4479}{0.7157} + \frac{0.0452}{(0.7157)^2}x + \frac{0.0452(0.5470)}{(0.7157)^3}x^2$$
$$+ \frac{0.0452(0.5470)^2}{(0.7157)^4}x^3 + \cdots.$$

And finally,

$$f_3(x) = f\{f_2(x)\} = \frac{0.4825 - 0.0717\left\{\dfrac{0.4479 - 0.2792x}{0.7157 - 0.5470x}\right\}}{1 - 0.5893\left\{\dfrac{0.4479 - 0.2792x}{0.7157 - 0.5470x}\right\}}$$

$$= \frac{(0.4825)(0.7157)-(0.4825)(0.5470x)-(0.0717)(0.4479)+(0.0717)(0.2792x)}{0.7157-0.5470x-(0.5893)(0.4479)+(0.5893)(0.2792x)}$$

or,

$$f_3(x) = \frac{0.3132 - 0.2439x}{0.4518 - 0.3825x}.$$

As with $f_2(x)$, $f_3(x)$ is of the general form $\frac{d-ex}{g-hx}$, with the values of d, e, g, and h now equal to $d = 0.3132$, $e = 0.2439$, $g = 0.4518$, and $h = 0.3825$. Since $hd - eg = (0.3825)(0.3132) - (0.2439)(0.4518) = 0.0096$, we have

$$f_3(x) = \frac{0.3132}{0.4518} + \frac{0.0096}{(0.4518)^2}x + \frac{0.0096(0.3825)}{(0.4518)^3}x^2$$
$$+ \frac{0.0096(0.3825)^2}{(0.4518)^4}x^3 + \cdots.$$

We can now directly read off the numerical answers to our questions from the power series forms of $f_2(x)$ and $f_3(x)$. Specifically,

the probability there are two male descendents in the second generation via a male-only sequence of ancestors is $\frac{0.0452(0.5470)}{(0.7157)^3}$ = 0.0674; the probability there are four male descendents in the second generation via a male-only sequence of ancestors is $\frac{0.0452(0.5470)^3}{(0.7157)^5}$ = 0.0394; the probability there are six male descendents in the third generation via a male-only sequence of ancestors is $\frac{0.0096(0.3825)^5}{(0.4518)^7}$ = 0.0205;

Okay, let's now suppose that we know none of the above; how can we simulate the process of family name propagation through time from one generation to the next, father to son? The code boom.m (which in turns calls the subroutine offspring.m) does that for the three generations specifically asked about in the original problem statement. Here's how those codes work.

The variable gen1 and the vectors gen2 and gen3 in boom.m have values that are equal to the number of males in the first, second, and third generations, respectively. gen1 is set equal to the number of sons born to our initial lone man; that number is selected from the integers 0 through 7 using Lotka's probabilities. The code boom.m will, in fact, be requesting such random integer numbers at several different places, and so in the code I've elected to write that part of the simulation as the subroutine offspring.m (this is the only problem solution in the book that uses subroutine calls). So, what happens in boom.m, in lines 01 through 07, is first the Lotka probabilities p0 through p6 are calculated, and then, using those probabilities, the variables zero, one, two, three, four, five, and six are calculated in lines 08 through 14. For example, zero is the probability a man has at most zero sons (i.e., has zero sons), while three is the probability a man has at most three sons.

Let me now discuss how the subroutine function offspring.m works. The syntax used to define a MATLAB function that has both input and output arguments is shown in line 01: first, the word function is followed by a list of output arguments (in offspring.m there is just one, sons), then there is an = symbol, followed by the subroutine function name and a list of input arguments. The input arguments are the values of zero, one, ..., six that have already been calculated in boom.m. The logic of offspring.m should now be clear: in line 02 the variable luck is given a random value from 0 to 1 and then, depending on what that value is, the output variable sons receives an integer value from 0 to 7 with the Lotka probability distribution. What follows next is how boom.m itself works.

Line 15 initializes the three-element vector answer to zero, where answer(1), answer(2), and answer(3) will, by the end of the simulation, be the number of simulations that result in two and four males in the second generation and six males in the third generation, respectively. Line 16 sets the value of the variable total, which is the total number of simulations performed by the code (I will give you the results produced by boom.m for total equal to 10,000 and for total equal to 10,0000). Lines

17 and 40 define the main for/end loop that cycles the code through all total simulations. Lines 18 and 19 initialize the vectors gen2 and gen3 to all zeros. gen2 is a seven-element vector where gen2(j) equals the number of second generation males due to the *j*th male in the first generation. gen3 is a forty-nine-element vector, where gen3(k) equals the number of third-generation males due to the *k*th male in the second generation; i.e., there could be as many as forty nine second-generation males. Now, with all of this preliminary setup work done, the actual construction of a random family tree begins.

Line 20 sets gen1 equal to an integer between 0 and 7 inclusive, via a call to the subroutine function offspring.m; gen1 receives the value of offspring.m's output variable sons, and this sets the number of males in the first generation.

Each of the males in the first generation, a total of gen1, then produces *his* sons (the second generation) with the loop in lines 21 through 23. Notice carefully that if (as is quite likely) gen1 < 7, then the first gen1 cells of the vector gen2 will each be a number from 0 to 7, and the remaining cells of gen2 will remain at their initialized values of zero. Notice, too, that if gen1 = 0, i.e., if the initial lone male of the 0th generation fails to have any sons at all, then the loop1 loop is automatically skipped, and there are no second-generation males (just we would expect!).

And finally, the loop of lines 25 through 30 generates the third generation. Since there could be 49 males in the second generation (because each of the seven first-generation sons could himself, have seven sons) there could be as many as 343 males in the third generation. That is, the vector gen3 has forty nine elements, with each cell holding a number from 0 to 7. What boom.m does in lines 24 through 30 is to look at each of the gen1 total cells in gen2 and generate as many Lotka-prescribed sons as the value specified by each gen2 cell; those numbers are then stored in the cells of gen3, using the variable index—initialized to 1 in line 24—as a pointer to the current cell of gen3 ready to receive a value of sons. Lines 31 through 39 then simply sum the elements of the gen2 and gen3 vectors and check to see if the desired conditions are met and, if so, update the elements in answer.

boom.m

```
01      p0 = 0.4825;
02      p1 = 0.2126;
03      p2 = p1*0.5893;
04      p3 = p2*0.5893;
05      p4 = p3*0.5893;
06      p5 = p4*0.5893;
07      p6 = p5*0.5893;
08      zero = p0;
09      one = zero + p1;
10      two = one + p2;
11      three = two + p3;
12      four = three + p4;
13      five = four + p5;
14      six = five + p6;
15      answer = zeros(1,3);
16      total = 10000;
17      for loop = 1:total
18          gen2 = zeros(1,7);
19          gen3 = zeros(1,49);
20          gen1 = offspring(zero,one,two,three,four,five,six);
21          for loop1 = 1:gen1
22              gen2(loop1) = offspring(zero,one,two,three,four,
                    five,six);
23          end
24          index = 1;
25          for loop2 = 1:gen1
26              for loop3 = 1:gen2(loop2)
27                  gen3(index) = offspring(zero,one,two,three,four,
                        five,six);
28                  index = index + 1;
29              end
30          end
31          n = sum(gen2);
32          if n == 2
33              answer(1) = answer(1) + 1;
```

(continued)

(continued)

```
34          elseif n == 4
35              answer(2) = answer(2) + 1;
36          end
37          if sum(gen3) == 6
38              answer(3) = answer(3) + 1;
39          end
40      end
41      answer/total
```

---

offspring.m

```
01      function sons = offspring(zero,one,two,three,four,five,six)
02      luck = rand;
03      if luck < zero
04          sons = 0;
05      elseif luck < one
06          sons = 1;
07      elseif luck < two
08          sons = 2;
09      elseif luck < three
10          sons = 3;
11      elseif luck < four
12          sons = 4;
13      elseif luck < five
14          sons = 5;
15      elseif luck < six
16          sons = 6;
17      else
18          sons = 7;
19      end
```

When run three times for total $= 10,000$, and then three more times for total $= 100,000$, the simulation produced the following estimates for the probabilities asked for:

Number of simulations = 10,000

2 males in second generation	4 males in second generation	6 males in third generation
0.0706	0.0406	0.0226
0.0733	0.0396	0.0205
0.0727	0.0419	0.0226

Number of simulations = 100,000

2 males in second generation	4 males in second generation	6 males in third generation
0.0688	0.0387	0.0206
0.0679	0.0409	0.0214
0.0680	0.0406	0.0212

The above probabilities are quite stable from run to run, for both values of total, and all are pretty close to the theoretical values computed earlier (although, as you would expect, the estimates for 100,000 simulations are the better ones). Remember, however, that the theory assumed each male could have *any* number of sons.

References and Notes

1. David Hawkins and S. Ulam, "Theory of Multiplicative Processes" (Los Alamos Scientific Laboratory Report LA-171, November 14, 1944, declassified in 1956). This report is reproduced in *Analogies Between Analogies* (*The Mathematical Reports of S. M. Ulam and His Los Alamos Collaborators*), A. R. Bednarek and Francoise Ulam, editors (Berkeley and Los Angeles: University of California Press, 1990, pp. 1–15). The word *multiplicative* has been replaced in the modern probability literature with *branching*.

APPENDIX 1

One Way to Guess on a Test

The code test.m simulates, a million times, the following method for guessing on the matching test described in the introduction: the student assigns to each president a term, selected at random each time, from the entire term list. The code test.m is similar to guess.m (the first four and the last nine lines are, in fact, identical), with only the details of an individual test differing. Line 05 defines the vector term with M elements (all initialized to zero). Lines 06 through 08 then randomly set the M elements of term, each, to one of the integers 1 to 24 (the MATLAB command ceil is a "round up" command). That's it! When run for the values of M used in the introduction (5, 10, 24, and 43), test.m gave the following values for the average number of correct pairings: 1.001114, 0.999015, 1.000011, and 1.000074, respectively. These results strongly suggest that the average number of correct pairings with this method of guessing is one, independent of the value of M.

---

test.m

```
01    M = 24;
02    totalcorrect = 0;
03    for k = 1:1000000
```

(continued)

(continued)

```
04        correct = 0;
05        term = zeros(1,M);
06        for j = 1:M
07              term(j) = ceil(M*rand);
08        end
09        for j = 1:M
10              if term(j) == j
11                    correct = correct + 1;
12              end
13        end
14        totalcorrect = totalcorrect + 1;
15     end
16     totalcorrect/1000000
```

Appendix 2

An Example of Variance Reduction in the Monte Carlo Method

The material in this appendix, somewhat more technical than most of the rest of the book, is an elaboration on an issue raised in the introduction: How good are the results of a Monte Carlo simulation? This question is a very deep one, and one could, if one wished, spend the rest of one's career pursuing the answer. Because I don't wish to do so, this appendix will have to suffice. Still, if the discussion that follows prompts you to look further into the matter on your own,[1] that's good, and I applaud you. You'll be a better analyst for the effort. For most of us, however, it will be sufficient (as I state in the introduction) to be convinced that running a simulation 10,000 times gives pretty nearly the same answer(s) as does running it 1,000,000 times.[2] But there are some very clever things one could do, if one wished, that are more subtle than the brute force "just simulate a lot" approach. What follows is an example (dating from 1956) of just one such possibility.

Looking back at Figure 1 in the introduction, which shows the geometry of a Monte Carlo estimation for π, you can see (once you remember the area interpretation of integration, and that the equation of the circle with radius 1 centered on the origin is $y^2 = 1 - x^2$) that what we are trying to do is equivalent to evaluating the integral $\int_0^1 \sqrt{1-x^2}\,dx$. In the introduction I asked you to imagine a random throwing of darts at the unit square in which a quarter-circle is inscribed (mathematically described by $y = f(x) = \sqrt{1-x^2}$), and we then estimated π by looking at how many darts landed inside and how many outside the circle. To do this, we used two random numbers per dart to establish the landing coordinates for each dart. To illustrate one way to decrease the Monte Carlo estimation error for π, as compared to the brute force, throw-more-darts approach, let me now formulate the evauation of the above integral in a slightly different way.

Let's write the exact value of our integral as V, and the average value of $f(x)$ over the integration interval of 0 to 1 as F. Then, again remembering the area interpretation of integration, we have

$$F = \frac{V}{\text{Upper limit} - \text{Lower limit}} = \frac{V}{1-0} = V.$$

This incredibly benign-looking result says that, to estimate π (actually, $\frac{\pi}{4}$), all we need do is find the average value of $f(x)$. We can do that by simply taking a lot of random values of x, uniformly distributed over the interval 0 to 1, calculate the value of $f(x)$ for each random x, and then calculate the average of those values. Our estimate for π is then just four times this average. This is quite easy to do; the code average.m does the job, and the results for $N = 100, 10{,}000$, and $1{,}000{,}000$ random x's were $3.087\ldots$, $3.1246\ldots$, and $3.14077\ldots$, respectively. As N increases we see that the estimation error does indeed decrease (although, even for N as large as a million, the accuracy of the estimate for pi isn't really very impressive).

---

average.m

```
01    N=input('How many x-values?');
02    sum=0;
03    for loop=1:N
04         x=rand;
05         sum=sum+sqrt(1-x^2);
06    end
07    4*sum/N
```

In the special case of an integral with a constant integrand, this Monte Carlo approach is certain to produce the exact value of V for any value of N. (This is not true for the throwing darts version of Monte Carlo.) This is because, since F is $f(x)$ as $f(x)$ is constant, we have as before that

$$V = \int_0^1 f(x)\,dx = \int_0^1 F\,dx = F.$$

And, of course, the average of any number of values, all equal to F, is F, which we've just seen equals V. This behavior is a direct consequence of the fact that the integral's integrand doesn't vary (has zero variance) over the entire interval of integration. Returning to our original problem of evaluating the integral $\int_0^1 f(x)\,dx$, where again $f(x) = \sqrt{1-x^2}$, we can take advantage of the above observation by writing our integral in the alternative form

$$V = \int_0^1 f(x)\,dx = \int_0^1 \frac{1}{2} f(x)\,dx + \int_0^1 \frac{1}{2} f(x)\,dx$$

which seems, at first glance, to be a trivial thing to do. But notice that if in the second integral on the right we change variable to $u = 1 - x$, then $du = -dx$ and so

$$V = \frac{1}{2} \int_0^1 f(x)\,dx + \frac{1}{2} \int_1^0 f(1-u)(-du)$$

$$= \frac{1}{2} \int_0^1 f(x)\,dx + \frac{1}{2} \int_0^1 f(1-u)\,du$$

or, more interestingly,

$$V = \int_0^1 \frac{1}{2}[f(x) + f(1-x)]\,dx.$$

The reason for why this is interesting comes from the fact that our new integrand, $\dfrac{\sqrt{1-x^2} + \sqrt{1-(1-x)^2}}{2}$, is "more constant" over the integration interval 0 to 1 than is our original integrand of $\sqrt{1-x^2}$.

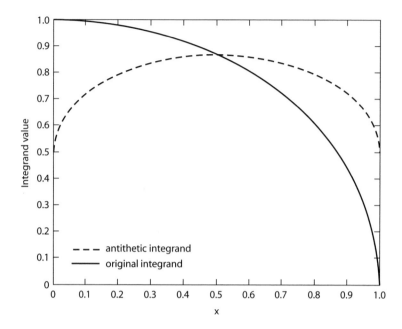

Figure A2.1. Our two integrands.

That is, our new integrand, when evaluated at random values of x to support a calculation of F, will exhibit a reduced variance. This is the result of $f(x)$ and $f(1-x)$ varying in opposite senses as x varies from 0 to 1 (they have, as a mathematician would put it, negative correlation), resulting in a reduced variation of their sum. And that is due to the fact that, in our problem, $f(x)$ is a monotonic function. If $f(x)$ were not monotonic than this method will be greatly, if not completely, reduced in effectiveness.[3] Figure A2.1 shows superimposed plots of the two integrands (statisticians call $1-x$ the *antithetic* of x, and so $\frac{\sqrt{1-x^2}+\sqrt{1-(1-x)^2}}{2}$ is the *antithetic integrand*.) And, in fact, when line 05 of average.m is replaced with

    sum = sum + (sqrt(1−x^2) + sqrt(1−(1−x)^2))/2

the results for $N = 100$, 10,000, and 1,000,000 random x's were (using the same random x's that were used in the first executions of average.m): 3.16005..., 3.136888..., and 3.1417324..., respectively. These estimates have, for a given N, less error than before.

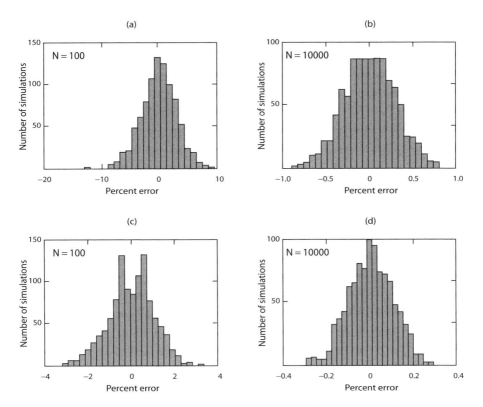

Figure A2.2. The error distribution in estimating using the original and the antithetic integrands.
(a) Original integrand, with $N = 100$. (b) Original integrand, with $N = 10,000$. (c) Antithetic integrand, with $N = 100$. (d) Antithetic integrand, with $N = 10,000$.

To really get a proper idea of what this reformulation of the problem has accomplished, however, we should do what was done in Figure 2 of the introduction, i.e., plot the error distribution in the estimation of pi for each of the two integrands. This is done in Figure A2.2, where 1,000 executions of average.m were performed for each of our two integrands for the two cases of, first, 100 random x's and then, second, 10,000 random x's. Histograms were then created from the results, just as was done in pierror.m. For $N = 100$ we can say, with high confidence, that the error made in estimating pi with the original integrand is no more than about $\pm 10\%$, while using the antithetic integrand reduces the width of the error interval to about ±3%. For $N = 10,000$, the error intervals are ±1% (original integrand) and ±0.3% (antithetic integrand). That is, for both $N = 100$ and $N = 10,000$, the antithetic integrand has reduced the width of the error interval by a factor of about three.

References and Notes

1. See, for example, Reuven Y. Rubinstein, *Simulation and The Monte Carlo Method* (New York: John Wiley & Sons, 1981), particularly Chapter 4, or J. M. Hammersley and D. C. Handscomb, *Monte Carlo Methods* (New York: Methuen & Co., 1964), particularly Chapter 5.

2. A wonderfully funny story, based on having a (much) too small sample size, is told by E. Bright Wilson, a professor of chemistry at Harvard, in his book, *An Introduction to Scientific Research* (New York: McGraw-Hill, 1952, p. 46): In the past there have been many ludicrous cases of conclusions drawn from an insufficient number of experiments. A story is told of an investigation in which chickens were subjected to a certain treatment. It was then reported that $33\frac{1}{3}$% of the chickens recovered, $33\frac{1}{3}$% died, and no conclusion could be drawn from the other $33\frac{1}{3}$% because that one ran away!

3. As Rubinstein (note 1) writes (p. 121), "Variance reduction can be viewed as a means to use *known information* [my emphasis] about the problem. In fact, if nothing is known about the problem, variance reduction cannot be achieved. At the other extreme, that is, complete knowledge, the variance is equal to zero and there is no need for simulation." "Variance reduction cannot be obtained from nothing; it is merely a way of not wasting information." In the example I use in the text, the "known information" is, of course, the monotonicity of the integrand.

APPENDIX 3

Random Harmonic Series

The code rhs.m produces a histogram of 50,000 values of the partial sums (first 100 terms) of $\sum_{k=1}^{\infty} \frac{t_k}{k}$, where $t_k = -1$ or $+1$ with equal probability. This histogram, shown in Figure A3.1, suggests that the probability is very small that the absolute value of a partial sum exceeds 4. But, remember, there is in fact no upper bound on the sum of the RHS; with an extremely small probability a sum could exceed any given finite value. The only "MATLABy" thing about rhs.m is the very last command, the highly useful hist. (I used it in the introduction to generate Figure 2, and in Appendix 2 to produce Figure A2.2.) That command produces a 50-bin histogram of the 50,000 values of sum stored in the vector sums. If your favorite language doesn't have a similar command, then you'll have to write some additional code.

---

rhs.m

```
01      sums = zeros(1,50000);
02      for loop = 1:50000
03          sum = 0;
04          for k = 1:100
05              t = rand;
```

(continued)

(continued)

```
06                    if t < 0.5
07                        t = 1;
08                    else
09                        t = −1;
10                    end
11                    sum = sum + t/k;
12                end
13                sums(loop) = sum;
14            end
15        hist(sums,50)
```

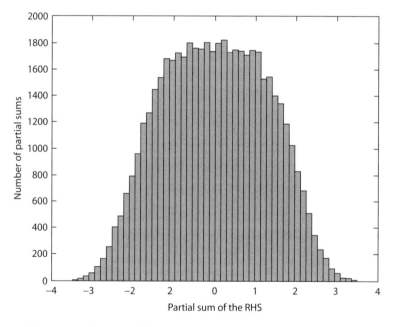

Figure A3.1. How the RHS behaves.

APPENDIX 4

Solving Montmort's Problem by Recursion

An alternative, everyday formulation of the President/term match-ing problem discussed in the introduction is the so-called Dinner Problem:

> Suppose n people are invited to a dinner party. Seats are as-signed and a name card made for each guest. However, floral arrangements on the table unexpectedly obscure the name cards. When the n guests arrive, they seat themselves randomly [for a more precise explanation of just what "random seating" means, read on]. What is the probability that no guest sits in his or her assigned seat?

This is vos Savant's president/term matching problem with the presi-dents interpreted as the dinner guests and the terms interpreted as the table seats.

Now, to be perfectly clear about the exact meaning of random seating, imagine that nobody is seated until after all n people have arrived. For a while, all n guests mill around munching crackers and dip, eating peanuts, and chitchatting with one another. As they mix and move about and around the table, we imagine that there is always some seat that each person happens to be nearer to than is anyone else.

Then, as the clock strikes six o'clock in the evening, all simultaneously sit in the seat that they are nearest to at that moment.

It is a curious fact that, to compute the probability that none has sat in his or her assigned seat, a great simplification occurs in the analysis if we generalize the problem and answer what appears to be a *harder* question! Our original problem will be a special case of the harder problem. (This is a clever idea that is one for any analyst to keep in his or her bag of tricks; it can work in a variety of situations, not just the one we are considering here.) Specifically, let's imagine that when each guest received the invitation they were told that if, at the last minute, they couldn't attend, it would be okay to send a substitute in their place. Thus, there will always be n people at the dinner, but not all will necessarily be original invitees; some may be what we'll call *substitutes*. We imagine that these substitutes arrive without giving prior notification to the dinner host, and so there is no chance there will be name cards on the table for them. This is, in fact, a uniquely defining characteristic property of a substitute—it is impossible for a substitute to sit at an assigned seat.

Let's now make the following definition:

$p_{n,k}$ = probability that no one sits at an assigned seat when k of
the n people in attendance are substitutes, $n \geq k \geq 0$.

For example, if $n = 1$, then the one person who is at the dinner party is either an invitee ($k = 0$) and so must sit at the one and only seat available, which is, of course, assigned to her, or is a substitute ($k = 1$) and so must sit at the one and only seat available, which is, of course, not assigned to her. Thus, $p_{1,0} = 0$ and $p_{1,1} = 1$. Suppose now that $n > 1$ and that there are no substitutes ($k = 0$). The value of $p_{n,0}$ is the answer to the original Montmort matching problem, and we can express it recursively as follows.[1]

Fasten your attention on one (any one) of the n invitees. The probability she will sit in one of the $n - 1$ seats assigned to someone else is $(n - 1)/n$. This means the person who was assigned that seat *must* sit in an unassigned seat. That is, it is impossible for that displaced person to sit at the seat assigned to her, which you'll recall is the distinguishing characteristic of a substitute. So, with probability

$(n-1)/n$ we have $n-1$ original invitees plus one "substitute" sitting down around the table. The probability none of them sit at an assigned seat is, by definition, $p_{n-1,1}$. Thus,

$$p_{n,0} = \frac{n-1}{n} p_{n-1,1}, \quad n > 1.$$

This says, for $n = 2$, for example, that

$$p_{2,0} = \frac{2-1}{2} p_{1,1} = \frac{1}{2} \times 1 = \frac{1}{2},$$

a result clearly true by inspection. Our boxed expression isn't quite enough to allow the calculation of $p_{n,0}$ for $n > 2$, however. We need something more.

To get that something more, imagine now that there is at least one substitute ($k \geq 1$) at the party, and fasten your attention on one (any one) of those substitutes. There is, of course, no possibility that this substitute will sit at an assigned seat. With probability k/n she will sit in a seat assigned for one of the k original invitees who instead sent substitutes (including herself), and with probability $(n-k)/n$ she will sit in a seat assigned to one of the original invitees who in fact actually came to the party. Notice carefully that in this second case the original invitee whose seat our substitute has taken now has no chance herself of sitting at her assigned seat. That is, in this case the unseated original invitee has taken on the uniquely defining characteristic of a substitute, i.e., one for whom there is zero chance of sitting at an assigned seat. In the first case (probability k/n) with our original substitute now seated, there are $n-1$ other people (of whom $k-1$ are substitutes) to be seated. The probability none of them sits at an assigned seat is, by definition, $p_{n-1,k-1}$. In the second case (probability $(n-k)/n$) with our original substitute seated, there are $n-1$ people and k (not $k-1$) substitutes left. The probability none of them sits at an assigned seat is, by definition, $p_{n-1,k}$. Together, all this gives us a second recursion:

$$p_{n,k} = \frac{k}{n}p_{n-1,k-1} + \frac{n-k}{n}p_{n-1,k}, \quad n>1, \quad n \geq k>0.$$

Now we can calculate $p_{n,0}$ for all $n > 2$.

For example, to calculate $p_{3,0}$ we need to know the value of $p_{2,1}$ and so we write

$$p_{2,1} = \frac{1}{2} \times p_{1,0} + \frac{1}{2} \times p_{1,1} = \frac{1}{2} \times 0 + \frac{1}{2} \times 1 = \frac{1}{2}$$

and then

$$p_{3,0} = \frac{3-1}{3}p_{2,1} = \frac{2}{3} \times \frac{1}{2} = \frac{1}{3} = 0.3333\cdots.$$

These recursive calculations for $p_{n,0}$ very quickly become ever more tedious as n increases, and so we turn to MATLAB for help. The code dinner.m computes the values for $p_{n,0}$ for $2 \leq n \leq 20$, and the table immediately following shows how very quickly those values converge to $e^{-1} = 0.36787944117\cdots$. Indeed, $p_{14,0}$ agrees with e^{-1} out to the 11th digit.

```
dinner.m
01    pkzero = zeros(1,20);
02    pkone = zeros(1,20);
03    pkone(1) = 1;
04    for n = 2:20
05          pkzero(n) = ((n−1)/n)*pkone(n−1);
06          pkzero(n)
07          pkone(n) = (1/n)*pkzero(n−1) + ((n−1)/n)*pkone(n−1);
08    end
```

To understand the operation of dinner.m, think of the values of $p_{n,k}$ as numbers on the lattice points of a two-dimensional array, with n and k on the horizontal and vertical axes, respectively. Then pkzero

n	$p_{n,0}$
2	0.5
3	0.3333 $\cdots$
4	0.375 $\cdots$
5	0.3666 $\cdots$
6	0.3680555 $\cdots$
7	0.367857142857 $\cdots$
8	0.367881944444 $\cdots$
9	0.367879188712 $\cdots$
10	0.367879464285 $\cdots$
11	0.367879439233 $\cdots$
12	0.367879441321 $\cdots$
13	0.367879441160 $\cdots$
14	0.367879441172 $\cdots$

and pkone are row vectors equivalent to the lattice points along the horizontal rows in that array associated with the $k = 0$ (the n-axis) and $k = 1$ cases, respectively. The value of $p_{1,0}$ is automatically set to zero in line 01, and line 03 sets the value of $p_{1,1}$ to one. Then, lines 04 through 08 simply implement the two boxed recursion equations.

The recursive method discussed here is a powerful one, and in his paper Brawner applies it to a related problem that he calls the Dancing Problem:

> Suppose n married couples ($2n$ people) are invited to a party. Dance partners are chosen at random, without regard to gender. What is the probability that nobody will be paired with his or her spouse?

Here the recursive approach, interestingly, converges very slowly, but still fast enough that Brawner could conjecture that the limiting value of the probability in question, as $n \to \infty$, is $1/\sqrt{e} = 0.60653\cdots$. His conjecture was soon shown to be correct, using a non-recursive approach.[2] The reason for why the dancing party problem, which at first glance might seem to simply be the dinner problem in disguise but instead has a quite different answer, is that while in the dancing problem any person can be paired with any other person, in the dinner problem nobody is going to mistake a person for a seat and sit on them!

References and Notes

1. This problem, and the recursion equations developed in this appendix, find their inspiration in a paper by James N. Brawner, "Dinner, Dancing, and Tennis, Anyone?" (*Mathematics Magazine*, February 2000, pp. 29–36).

2. Barbara H. Margolius, "Avoiding Your Spouse at a Bridge Party" (*Mathematics Magazine*, February 2001, pp. 33–41).

APPENDIX 5

An Illustration of the Inclusion-Exclusion Principle

To appreciate the value of the inclusion-exclusion principle mentioned in note 4 of the introduction, consider the following problem from the excellent textbook by Saeed Ghahramani, *Fundamentals of Probability*, (Upper Saddle River, N. J: Prentice-Hall, 1996, p. 69):

> From a faculty of six professors, six associate professors, 10 assistant assistant professors, and 12 instructors, a committee of size 6 is formed randomly. What is the probability that there is at least one person from each rank on the committee?
> Hint: Be careful, the answer is not

$$\frac{\binom{6}{1}\binom{6}{1}\binom{10}{1}\binom{12}{1}\binom{30}{2}}{\binom{34}{6}} = 1.397.$$

Professor Ghahramani then suggests the inclusion-exclusion principle as the proper tool with which to answer the question.

The incorrect answer above is almost certainly what most beginning probability students would indeed write, at least until they evaluate the expression on the left and get that "probability" on the right that is greater than one (a really big clue that something isn't quite right)! The first issue for us to dispose of, of course, is that of explaining just why the expression on the left leads to such an obviously impossible answer. The explanation is actually pretty simple. First, the denominator itself *is* correct, as it is the number of ways to randomly select six people from thirty four, i.e., $\binom{34}{6}$ is the total number of distinct committees. The individual binomial coefficients in the numerator are, respectively, from left to right, the number of ways to select one professor from six, one associate professor from six, one assistant professor from ten, one instructor from twelve (at this point we have one person from each rank), and then any two from the thirty people who haven't yet been selected. It is an easy jump from this to the conclusion that the product of these individual numbers is the total number of committees with the desired structure, but as we now know, that is a faulty jump. The problem with the reasoning is that it counts many possible committees more than once. And that's why the incorrect expression is larger than one. For example, suppose we number the thirty four people as follows:

[1 to 6]	[7 to 12]	[13 to 22]	[22 to 34]
full	associate	assistant	instructor

Then one possible committee from the incorrect numerator is, from left to right, 1, 7, 13, 23, 5, 8. But the incorrect numerator also generates the committee 5, 8, 13, 23, 1, 7—which is, of course, the same committee. What we need is a way to count all the distinct committees that have the desired structure without repetition. That's what the inclusion-exclusion principle allows us to do.

So, you may now be asking, just what is the inclusion-exclusion principle? Suppose $E_1, E_2, \ldots, E_n$ are n events defined on a common sample space (the collection of all possible outcomes when some experiment is performed; e.g., in our problem, the experiment is the selection of six people from thirty four to form a committee). Then the probability that at least one of these events occurs simultaneously is given by (the $\cup$ symbol is the mathematician's notation for the logical inclusive-OR, equivalent to the plus sign commonly used by digital circuit design engineers)[1]

$$\text{Prob}(E_1 \cup E_2 \cup \cdots \cup E_n) = \sum_{i=1}^{n} \text{Prob}(E_i) - \sum_{i=1}^{n} \sum_{j=i+1}^{n} \text{Prob}(E_i E_j)$$

$$+ \sum_{i=1}^{n} \sum_{j=i+1}^{n} \sum_{k=j+1}^{n} \text{Prob}(E_i E_j E_k) - \cdots.$$

For $n = 3$, for example,

$$\text{Prob}(E_1 \cup E_2 \cup E_3) = \text{Prob}(E_1) + \text{Prob}(E_2) + \text{Prob}(E_3)$$
$$- \text{Prob}(E_1 E_2) - \text{Prob}(E_1 E_3) - \text{Prob}(E_2 E_3)$$
$$+ \text{Prob}(E_1 E_2 E_3).$$

For our committee problem, define four elementary events as follows:

$E_1 = \{\text{"no full professors on the committee"}\};$
$E_2 = \{\text{"no associate professors on the committee"}\};$
$E_3 = \{\text{"no assistant professors on the committee"}\};$
$E_4 = \{\text{"no instructors on the committee"}\}.$

Then, the compound event $E_1 \cup E_2 \cup E_3 \cup E_4$ is the event that at least one of E_1, E_2, E_3, E_4 occur simultaneously, and so $\text{Prob}(E_1 \cup E_2 \cup E_3 \cup E_4)$ is the probability that at least one of the four elementary events occurs. Thus, $1 - \text{Prob}(E_1 \cup E_2 \cup E_3 \cup E_4)$ is the probability that none of the four elementary events occurs, which is of course the probability that there is at least one person of each rank on the committee—which is just what we want to calculate. But $1 - \text{Prob}(E_1 \cup E_2 \cup E_3 \cup E_4)$ is, by the inclusion-exclusion principle, given by

$$1 - \text{Prob}(E_1) - \text{Prob}(E_2) - \text{Prob}(E_3) - \text{Prob}(E_4) + \text{Prob}(E_1 E_2)$$
$$+ \text{Prob}(E_1 E_3) + \text{Prob}(E_1 E_4) + \text{Prob}(E_2 E_3) + \text{Prob}(E_2 E_4)$$
$$+ \text{Prob}(E_3 E_4) - \text{Prob}(E_1 E_2 E_3) - \text{Prob}(E_1 E_2 E_4)$$
$$- \text{Prob}(E_1 E_3 E_4) - \text{Prob}(E_2 E_3 E_4) + \text{Prob}(E_1 E_2 E_3 E_4).$$

Each of these terms is easily written down by inspection. For example,

$$\text{Prob}(E_1) = \frac{\binom{28}{6}}{\binom{34}{6}} = \text{Prob}(E_2)$$

and

$$\text{Prob}(E_1 E_2) = \frac{\binom{22}{6}}{\binom{34}{6}}.$$

The first numerator follows by noticing that for E_1 to occur (no full professors selected), all six members of the committee come from the twenty eight people who are not full professors, and similarly for E_2. The second numerator follows by noticing that for E_1 *and* E_2 to both occur (no full and no associate professors selected), all six members of the committee come from the twenty two people who are neither full nor associate professors. And of course, $\text{Prob}(E_1 E_2 E_3 E_4) = 0$, because *every* person on the committee is from *some* rank! So, the total expression for the probability we are after is

$$1 + \frac{-\binom{28}{6} - \binom{28}{6} - \binom{24}{6} - \binom{22}{6} + \binom{22}{6} + \binom{18}{6}}{\binom{34}{6}}$$

$$+ \frac{\binom{16}{6} + \binom{18}{6} + \binom{16}{6} + \binom{12}{6} - \binom{12}{6} - \binom{10}{6}}{\binom{34}{6}}$$

$$+ \frac{-\binom{6}{6} - \binom{6}{6}}{\binom{34}{6}}$$

$$= 1 + \frac{-2\binom{28}{6} - \binom{24}{6} + 2\binom{18}{6} + 2\binom{16}{6} - \binom{10}{6} - 2}{\binom{34}{6}}$$

$$= 1 + \frac{-753{,}480 - 134{,}596 + 37{,}128 + 16{,}016 - 210 - 2}{1{,}344{,}904}$$

$$= 1 - \frac{835{,}144}{1{,}344{,}904} = 0.379.$$

Your confidence in this calculation would surely be enhanced if a Monte Carlo simulation agrees, and that is what the code committee.m confirms; after generating 100,000 random committees of six people selected from thirty four, it produced an estimate for the probability a committee has the desired structure of 0.3778, in pretty good agreement with theory. The logic of committee.m is straightforward. In line 01, the variable structure is initialized to zero; at the end of the simulation its value will be the number of committees (out of 100,000) that had the required membership. The loop defined by lines 02 and 27 perform each committee simulation. In line 03, count is initialized to zero: once a committee is formed, count will be incremented by one when a member puts a new rank in place for the first time. Line 04 generates a random permutation in the vector mix of the thirty four total people available, where the integers 1 through 34 are to be interpreted as we used them in the earlier part of this appendix, e.g., 1 through 6 are the full professors, and so on. Lines 05, 06, and 07 take the first six elements of mix and use them to form the six-element vector select, which is in fact the committee. All we need do now is check select to see if the committee has the desired structure.

committee.m

```
01      structure = 0;
02      for loop = 1:100000
03          count = 0;
04          mix = randperm(34);
05          for i = 1:6
06              select(i) = mix(i);
07          end
08          rank = ones(1,4);
09          for i = 1:6
10              if select(i) < 7
11                  count = count + rank(1);
12                  rank(1) = 0;
13              elseif select(i) > 6 & select(i) < 13
14                  count = count + rank(2);
15                  rank(2) = 0;
16              elseif select(i) > 12 & select(i) < 23
17                  count = count + rank(3);
18                  rank(3) = 0;
19              else
20                  count = count + rank(4);
21                  rank(4) = 0;
22              end
23          end
24          if count == 4
25              structure = structure + 1;
26          end
27      end
28      structure/100000
```

To start this checking, line 08 creates the four-element vector rank, with all its elements set equal to one. These elements, rank(1), rank(2), rank(3), and rank(4), are to be associated with full, associate, and assistant professors and instructors, respectively. Then, in the loop defined by lines 09 and 26, the code looks at each committee member in turn (that is, at select(i) as i runs from 1 to 6) and determines which rank that member has. Depending on that rank, count is incremented by the corresponding element of rank — the first time this happens, for each rank, that increment is one—and then the value of that element is set equal to zero, which means that any subsequent committee member with the same rank will increment count by zero, i.e., will have no affect on determining the final value of count. After all six committee members have been examined, the value of count will be 4 if and only if there is at least one committee member of each rank. Lines 24 through 26 increment structure by one if count equals 4, and then another simulated committee is formed. Line 28 gives committee.m's estimate of the probability a randomly generated committee has the desired structure.

References and Notes

1. An analytical proof of a generalized inclusion-exclusion principle for an arbitrary number of n events can be found in the book by Emanuel Parzen, *Modern Probability Theory and Its Applications* (New York: John Wiley & Sons, 1960, pp. 76–85). If you are familiar with the digital circuit logic design tool of Karnaugh maps (what a mathematician would call Venn diagrams), then it is easy to construct visual proofs of the inclusion-exclusion principle that are obvious by inspection for the $n = 2, 3,$ and 4 cases.

APPENDIX 6

Solutions to the Spin Game

To theoretically analyze the spin game described at the end of the introduction, start by defining P as the probability of the event that the player, starting with disk 1, wins. Then we can write, as follows, all the ways that event can happen:

(1) the pointer stops in area p_{11} n times in a row, where $n = 0, 1, 2, 3 \ldots$, and then the pointer stops in area x_1;

(2) the pointer stops in area p_{12}, followed by the event that, if the player starts with disk 2, he wins the game (i.e., the pointer stops in area x_1);

(3) the pointer stops in area p_{11} n times in a row, where $n = 1, 2, 3, \ldots$, and then (2) occurs.

If we write Q as the probability that the player wins if he starts with disk 2, then we can immediately write

$$P = (p_{11}^0 x_1 + p_{11} x_1 + p_{11}^2 x_1 + p_{11}^3 x_1 + \cdots) + (p_{12}Q)$$

$$+ (p_{11}p_{12}Q + p_{11}^2 p_{12}Q + p_{11}^3 p_{12}Q + \cdots)$$

$$= x_1(1 + p_{11} + p_{11}^2 + p_{11}^3 + \cdots) + p_{12}Q(1 + p_{11} + p_{11}^2 + p_{11}^3 + \cdots)$$

$$= (x_1 + p_{12}Q)(1 + p_{11} + p_{11}^2 + p_{11}^3 + \cdots) = \frac{x_1 + p_{12}Q}{1 - p_{11}}$$

$$= \frac{1 - p_{11} - p_{12} + p_{12}Q}{1 - p_{11}}.$$

We can find Q with the same sort of reasoning. That is, starting with disk 2, the player wins if:

(1)   the pointer stops in area p_{21}, followed by the event that, if the player starts with disk 1, he wins the game;

(2)   the pointer stops in area p_{22} n times in a row, where $n = 1, 2, 3, \ldots$, and then (1) occurs.

Thus,

$$Q = p_{21}P + p_{22}p_{21}P + p_{22}^2 p_{21}P + p_{22}^3 p_{21}P + \cdots$$
$$= p_{21}P(1 + p_{22} + p_{22}^2 + p_{22}^3 + \cdots) = \frac{p_{21}P}{1 - p_{22}}.$$

Inserting this result for Q into our earlier result for P, we have

$$P = \frac{1 - p_{11} - p_{12} + p_{12}\dfrac{p_{21}P}{1 - p_{22}}}{1 - p_{11}}$$

which is easily solved for P to give

$$P = \frac{(1 - p_{11} - p_{12})(1 - p_{22})}{(1 - p_{11})(1 - p_{12}) - p_{12}p_{21}}.$$

For the p_{ij} values given in the original problem statement, this says

$$P = \frac{(1 - 0.2 - 0.4)(1 - 0.35)}{(1 - 0.2)(1 - 0.4) - (0.4)(0.35)} = 0.5821.$$

To write a Monte Carlo simulation of this game, the flow diagram of Figure A6.1 illustrates the logic of determining the sequence of pointer spins for a single simulated game, from the first spin of the pointer of disk 1 until the game ends. The value of the variable disk is the current disk being used, and p(i,j) $= p_{ij}$. The code spin.m implements the logic

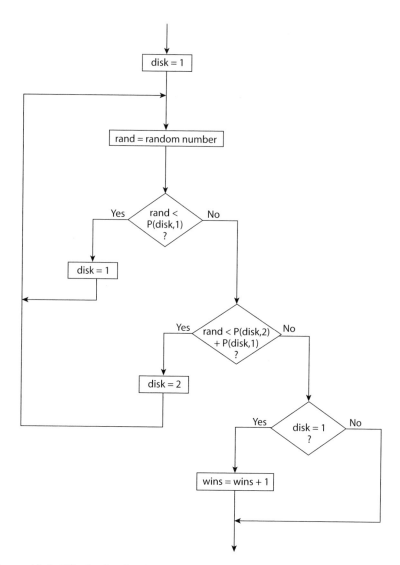

Figure A6.1. The logic of a simulated spin game.

of Figure A6.1. When run for ten thousand simulated games, the code produced an estimate of 0.5791 for P, while increasing the number of simulated games to ten million gave an estimate for P of 0.5818, in pretty good agreement with theory. Just in case you're curious, while the execution times of any simulation are of course dependent on the details of the computer used, my machine (with a 3-GHz Pentium 4 and 1 GByte of RAM) required 0.0073 seconds to simulate 10,000 games and 4.6 seconds to run through 10,000,000 games.

```
spin.m
01    p(1,1) = 0.3;p(1,2) = 0.4;p(2,1) = 0.3;p(2,2) = 0.35;
02    wins = 0;
03    for loop = 1:10000
04        keepgoing = 1;
05        disk = 1;
06        while keepgoing == 1
07            spinpointer = rand;
08            if spinpointer < p(disk,1)
09                disk = 1;
10            else
11                if spinpointer < p(disk,1) + p(disk,2)
12                    disk = 2;
13                else
14                    keepgoing = 0;
15                    if disk == 1
16                        wins = wins + 1;
17                    end
18                end
19            end
20        end
21    end
22    wins/10000
```

APPENDIX 7

How to Simulate Kelvin's Fair Coin with a Biased Coin

Look again at note 7 in the introduction for the motivation behind this appendix. Suppose the biased coin has probability p of showing heads, and so probability $1 - p$ of showing tails. Suppose further that we assume individual tosses of the coin are independent. Toss the coin twice. It shows HH with probability p^2, TT with probability $(1 - p)^2$, and either TH or HT with equal probability $p(1 - p)$. So, if TH appears, call it heads, and if HT appears, call it tails. If either HH or TT appears, ignore the result and toss twice again. The price we pay for making any coin into a fair coin by this method is time. That is, we will have to toss the coin multiple times to get a decision, while with a fair coin we get a decision on each toss. We can calculate this price as follows. Suppose we get a decision on the kth double toss. That means we did *not* get a decision on the previous $k - 1$ double tosses. Those $k - 1$ failures each occurred with probability $p^2 + (1 - p)^2$, while the terminating success occurred with probability $2p(1 - p)$. The probability of getting a decision on the kth double toss is therefore

$$[p^2 + (1 - p)^2]^{k-1} 2p(1 - p), \qquad k \geq 1.$$

The average number of double tosses to get a decision is then given by

$$\sum_{k=1}^{\infty} k[p^2 + (1-p)^2]^{k-1} 2p(1-p) = \frac{2p(1-p)}{c} \sum_{k=1}^{\infty} kc^k,$$

$$c = p^2 + (1-p)^2.$$

Define

$$S = \sum_{k=1}^{\infty} kc^k = c + 2c^2 + 3c^3 + 4c^4 + \cdots.$$

Thus,

$$cS = c^2 + 2c^3 + 3c^4 + \cdots$$

and so

$$S - cS = (1-c)S = c + c^2 + c^3 + c^4 + \cdots$$

or,

$$S = \frac{c(1 + c + c^2 + c^3 + c^4 + \cdots)}{1-c}.$$

Using the same trick to sum the geometric series $1 + c + c^2 + c^3 + c^4 + \cdots$ gives us the result

$$S = \frac{c \dfrac{1}{1-c}}{1-c} = \frac{c}{(1-c)^2}.$$

Thus, the average number of double tosses required to get a decision is

$$\frac{2p(1-p)}{c} \times \frac{c}{[1 - p^2 - (1-p)^2]^2} = \frac{2p(1-p)}{(1 - p^2 - 1 + 2p - p^2)^2}$$

$$= \frac{2p(1-p)}{(2p - 2p^2)^2} = \frac{2p(1-p)}{4p^2(1-p)^2} = \frac{1}{2p(1-p)}.$$

If, for example, $p = 0.4$ then the average number of double tosses required to get a decision is

$$\frac{1}{(0.8)(0.6)} = \frac{1}{0.48} = 2.0833,$$

i.e., on average one has to toss this biased coin four times to get a fair decision.

This problem also yields easily to a Monte Carlo simulation. The code kelvin.m does the job and, for $p = 0.4$, estimates (after 10,000 simulations) the average number of double tosses required to get a decision to be 2.085, which is in pretty good agreement with the theoretical answer. Increasing the number of simulations to one million (in line 03) gives the even better estimate of 2.0822. The operation of kelvin.m should be clear, but just to be sure, let me point out that toss1 and toss2 are the results of tossing the biased coin twice, with a value of 0 representing tails (T) and a value of 1 representing heads (H). We get a decision for either HT or TH, i.e., when toss1 + toss2 = 1, as determined in the if/end logic of lines 18 through 21. The while loop in lines 06 through 21 keeps tossing the coin until decision is set equal to 1 in line 19. (The three periods at the end of the first "line" of line 22 are, as in the code chess.m at the end of the introduction, MATLAB's way of continuing a line too long to fit the width of a page.)

```
kelvin.m
01    p = 0.4;
02    totalnumberofdoubletosses = 0;
03    for loop = 1:10000
04        numberofdoubletosses = 0;
05        decision = 0;
06        while decision == 0
07            if rand < p
08                toss1 = 1;
09            else
10                toss1 = 0;
```

(continued)

(continued)

```
11              end
12              if rand < p
13                  toss2 = 1;
14              else
15                  toss2 = 0;
16              end
17              numberofdoubletosses = numberofdoubletosses + 1;
18              if toss1 + toss2 == 1
19                  decision = 1;
20              end
21          end
22      totalnumberofdoubletosses = totalnumberofdoubletosses
                + ... numberofdoubletosses;
23  end
24  totalnumberofdoubletosses/10000
```

APPENDIX 8

How to Simulate an Exponential Random Variable

In Problem 15 (How Long Is the Wait to Get the Potato Salad?) it is necessary to generate random numbers that are not uniformly distributed from zero to one (which MATLAB's rand does for us) but rather are *exponentially* distributed from zero to infinity. That is, we need to simulate a random variable $\mathbf{T}$ whose values are described by the probability density function

$$f_{\mathbf{T}}(t) = \begin{array}{l} \lambda e^{-\lambda t}, \quad 0 \le t < \infty \\ 0, \qquad t < 0, \end{array}$$

where λ is an arbitrary positive constant. (In Problem 15, λ has an important physical interpretation.) MATLAB has no built-in function that does this, and so we must write some code ourselves that does the job. There are numerous approaches[1] that one could take to developing such a code, but the one I'll show you here is perhaps the easiest to understand as well as the simplest to implement.

Define $\mathbf{U}$ as MATLAB's uniform random variable, and let $F_{\mathbf{T}}(t)$ be the distribution function of $\mathbf{T}$, i.e.,

$$F_{\mathbf{T}}(t) = \text{probability } \mathbf{T} \le t = P(\mathbf{T} \le t).$$

As $F_{\mathbf{T}}(t)$ is a probability we know $0 \leq F_{\mathbf{T}}(t) \leq 1$, and since $\mathbf{U}$ is uniform from 0 to 1 we then immediately have

$$P(\mathbf{U} \leq F_{\mathbf{T}}(t)) = F_{\mathbf{T}}(t).$$

But this says

$$F_{\mathbf{T}}(t) = P(F_{\mathbf{T}}^{-1}(\mathbf{U}) \leq t).$$

But since $F_{\mathbf{T}}(t) = P(\mathbf{T} \leq t)$, we have our result: $\mathbf{T} = F_{\mathbf{T}}^{-1}(\mathbf{U})$.

For the particular case we are interested in, that of $\mathbf{T}$ as an exponential random variable, we have

$$F_{\mathbf{T}}(t) = \int_0^t \lambda e^{-\lambda s} ds = (-e^{-\lambda s} |_0^t = 1 - e^{-\lambda t}.$$

Now, by the very definition of an inverse function,

$$F_{\mathbf{T}}(F_{\mathbf{T}}^{-1}(t)) = t$$

and so

$$t = 1 - e^{-\lambda F_{\mathbf{T}}^{-1}(t)}.$$

This is easily solved for $F_{\mathbf{T}}^{-1}(t)$, i.e.,

$$F_{\mathbf{T}}^{-1}(t) = -\frac{1}{\lambda} \ln(1 - t),$$

and since we showed earlier that $F_{\mathbf{T}}^{-1}(\mathbf{U}) = \mathbf{T}$, we thus have

$$\mathbf{T} = -\frac{1}{\lambda} \ln(1 - \mathbf{U}).$$

We can simplify this just a bit by noticing that since $\mathbf{U}$ is uniform from 0 to 1, then clearly so is $1 - \mathbf{U}$. But that means we can replace $1 - \mathbf{U}$ with $\mathbf{U}$, thus saving a subtraction operation. Our final result is that

$$\mathbf{T} = -\frac{1}{\lambda}\ln(\mathbf{U}).$$

Thus, when in Problem 15 the need arises to generate a value for a random variable that is exponentially distributed from zero to infinity with parameter λ (defined as the variable lambda), the MATLAB code we will use is the single line

—log(rand)/lambda.

References and Notes

1. See, for example, John Dagpunar, *Principles of Random Variate Generation* (Oxford: Oxford University Press, 1988, pp. 90–93).

2. This approach to simulating a nonuniform random variable from a uniform one works only in the case where we can actually analytically invert $F_{\mathbf{T}}(t)$ to get $F_{\mathbf{T}}^{-1}(t)$. This is not always possible. For example, an analytical inversion can't be done in the case of a Gaussian (normal) random variable. Try it and see!

APPENDIX 9

Author-Created MATLAB m-files and Their Location in the Book

	Name	*Location*
1.	aandb.m	Problem 14 (solution)
2.	average.m	Appendix 2
3.	boom.m	Problem 21 (solution)
4.	broke.m	Problem 4 (solution)
5.	bus.m	Problem 17 (solution)
6.	car.m	Problem 3 (solution)
7.	chess.m	Introduction
8.	committee.m	Appendix 5
9.	deli.m	Problem 15 (solution)
10.	dinner.m	Appendix 4
11.	dish.m	Problem 1 (solution)
12.	easywalk.m	Problem 18 (solution)
13.	election.m	Problem 19 (solution)
14.	estimate.m	Problem 12 (solution)
15.	fb.m	Problem 9 (solution)
16.	floss.m	Problem 7 (solution)

	Name	Location
17.	gameb.m	Problem 14 (solution)
18.	gs.m	Problem 5 (solution)
19.	guess.m	Introduction
20.	jury.m	Problem 16 (solution)
21.	kelvin.m	Appendix 7
22.	malt.m	Problem 2 (solution)
23.	missing.m	Problem 11 (solution)
24.	mono.m	Introduction
25.	obtuse.m	Introduction
26.	obtuse1.m	Introduction
27.	offspring.m	Problem 21 (solution)
28.	optimal.m	Problem 20 (solution)
29.	patrol.m	Problem 13 (solution)
30.	pierror.m	Introduction
31.	rhs.m	Appendix 3
32.	rolls.m	Problem 8 (solution)
33.	smoker.m	Problem 7 (solution)
34.	smokerb.m	Problem 7 (solution)
35.	spin.m	Appendix 6
36.	steve.m	Problem 6 (solution)
37.	stopping.m	Problem 20 (solution)
38.	sylvester.m	Introduction
39.	test.m	Appendix 1
40.	umbrella.m	Problem 10 (solution)
41.	walk.m	Problem 18 (solution)

Glossary

Binomial coefficient: the number of ways to select k objects from n objects ($k \le n$) without regard to the order of selection; written as $\binom{n}{k}$ $= \frac{n!}{k!(n-k)!}$.

Concave region: a region that is not convex.

Confidence interval: the interval of values within which a parameter of a random variable X is declared to be with a specified probability; e.g., if the mean of X is estimated to be in the interval (a,b) with probability 0.95, then (a, b) is said to be a 95% confidence interval.

Convex region: a region is convex if and only if any two points in the region can be connected by a straight line that has all of its points within the region; a circular region is convex, while the same region with a hole punched in it is not.

Ensemble average: the average over an infinity of sample functions for a stochastic process at a given time; for a so-called stationary process, this average is independent of the given time.

Expected value: the average or mean value of a random variable.

Fair coin: a coin with equal probability (1/2) of showing heads or tails when flipped.

Histogram: a bar graph of the number of times the values of n measurements of a random variable fall into equal width subintervals (called bins) of an interval that spans the range from the smallest to the largest measurement; the sum of all the bin numbers equals n.

Lattice point: any point with integer-valued coordinates.

Mean value: the average or expected value of a random variable, calculated as follows: if X is a random variable with probability density function $f_X(x)$, then the mean value of X is $E(X) = m = \int_{-\infty}^{\infty} x f_X(x)\, dx$.

Permutation: a random rearrangement of a set of objects; there are $n!$ permutations of n objects.

Probability density function: any function $f_X(x)$ such that $f_X(x) \geq 0$ and $\int_{-\infty}^{\infty} f_X(x)\,dx = 1$. If X is a random variable, then $\int_a^b f_X(x)\,dx$ is the probability a measurement of X will be in the interval (a, b). If a probability density function of one random variable depends on the particular value of another random variable we then have a conditional probability density function, e.g., $f_X(x \mid Y = y)$, where the vertical bar $\mid$ is read as "given that." $f_X(x)$ can be calculated from $f_X(x \mid Y = y) = f_X(x \mid y)$ as $f_X(x) = \int_{-\infty}^{\infty} f_X(x \mid y)\,dy$.

Probability distribution function: the integral of a probability density function; i.e., $F_X(x) = \int_{-\infty}^{x} f_X(u)\,du$ is the probabilty a measurement of the random variable X will be in the interval $(-\infty, x)$. That is, $F_X(x)$ is the probability that $X \leq x$. Since a distribution is a probability, then $0 \leq F_X(x) \leq 1$. We can also write $f_X(x) = \frac{d}{dx} F_X(x)$.

Queue: a line of "customers" waiting for "service."

Random variable: a quantity that, in general, has a different value each time it is measured, with the probability a measurement is in an interval as the integral of a probability density function (over the interval).

Random walk: the steplike, time-discrete motion of a point, with the length or direction (or both) of each step described by a random variable; random walks can occur in spaces of dimension 1 (along a line), dimension 2 (in a plane), dimension 3 (in a space), or in higher dimensions as well.

Sample function: a particular observation over time of a stochastic process.

Sampling without replacement: the process of selecting, one at a time, a subset of objects from a population of objects and not returning each selected object to the population before making the next selection. As the population increases in size, distinguishing between sampling with or without replacement becomes unimportant.

Stochastic process: a random variable that is a function of time; if X is a stochastic process, then $X = X(t)$; more specifically, the probability density function of X is a function of t, i.e., $f_X(x) = f_X(x, t)$.

Uniform distribution: a probability density function that is constant over a finite length interval and zero everywhere else; commonly available in most computer-based random number generators over the interval 0 to 1.

Variance: a measure of the spread of a random variable, calculated as follows: if X is a random variable with probability density function $f_X(x)$ and mean value m, then the variance of X is $\sigma_X^2 = \int_{-\infty}^{\infty} (x - m)^2 f_X(x)\,dx$.

Vector: a one-dimensional sequence of n numbers is called a vector of length n; e.g., if the vector is called bob, then bob(1), bob(2), and bob(n) are the first, second, and nth (i.e., last) elements of the vector bob. A two-dimensional array of numbers is called a matrix; a matrix can be thought of as being constructed from either horizontal row vectors or vertical column vectors.

Acknowledgments

I had lots of help while writing this book. It was typed in Scientific Word (MacKichan Software), and all the computer codes were written in MATLAB (Mathworks). Both applications would have been science fantasy when I was in high school—I remember well being told to both take typing and to practice my slide-rule in those long-ago days!—and they have changed how technical writers now work. I can't imagine life without my SW and MATLAB. At Mathworks, in particular, I thank Courtney Esposito, who arranged for me to have the latest versions of MATLAB and Symbolic Toolbox.

At Princeton University Press, my long-time editor, Vickie Kearn, was ever helpful and supportive, and Debbie Tegarden made all the details of getting a submitted typescript through production click and clack rather than clash and clog. Marjorie Pannell, the book's copy editor, was a pleasure to work with.

The photograph of my high school math teacher Victor Hassing, to whom I've dedicated this book, was provided by his wife, Joyce Hassing. I am most grateful for her willingness to send a precious, unique photo to someone thousands of miles away on the faith of a telephone call out of the blue from a stranger.

My wife, Patricia Ann, made it all worth doing, even when I was faced with exploding hard drives, accidentally deleted text files, and evaporating backup disks. As I would sit in my study after each such disaster, wondering if I could possibly find the strength to go on, she would tiptoe up behind me, gently massage my shoulders, and then softly purr into an ear, "Better get writing again, sweetie! I just booked us on a trip to see our moms and charged the plane tickets to your credit card." *That* got me going!

Paul J. Nahin
Lee, New Hampshire
September 2007

Index

Also by Paul J. Nahin

Oliver Heaviside (1988, 2002)

Time Machines (1993, 1999)

The Science of Radio (1996, 2001)

An Imaginary Tale (1998, 2007)

Duelling Idiots (2002, 2002)

When Least Is Best (2004, 2007)

Dr. Euler's Fabulous Formula (2006)

Chases and Escapes (2007)